Building Bridges
Between Theory and Practice

Building Bridges

Between Theory and Practice

David Blockley
University of Bristol, UK

NEW JERSEY • LONDON • SINGAPORE • BEIJING • SHANGHAI • HONG KONG • TAIPEI • CHENNAI • TOKYO

Published by

World Scientific Publishing Europe Ltd.
57 Shelton Street, Covent Garden, London WC2H 9HE
Head office: 5 Toh Tuck Link, Singapore 596224
USA office: 27 Warren Street, Suite 401-402, Hackensack, NJ 07601

Library of Congress Cataloging-in-Publication Data
Names: Blockley, David, author.
Title: Building bridges : between theory and practice / David Blockley, University of Bristol, UK.
Description: New Jersey ; London : World Scientific, [2020] | Includes bibliographical references and index.
Identifiers: LCCN 2019026446 | ISBN 9781786347626 (hardcover) | ISBN 9781786347633 (ebook)
Subjects: LCSH: Failure analysis (Engineering) | Risk assessment. | Structural failures--Case studies.
Classification: LCC TA169.5 .B56 2020 | DDC 624.2/5--dc23
LC record available at https://lccn.loc.gov/2019026446

British Library Cataloguing-in-Publication Data
A catalogue record for this book is available from the British Library.

For any available supplementary material, please visit
https://www.worldscientific.com/worldscibooks/10.1142/Q0226#t=suppl

Desk Editors: Aanand Jayaraman/Jennifer Brough/Shi Ying Koe

Typeset by Stallion Press
Email: enquiries@stallionpress.com

Preface

How many deaths will it take till he knows that too many people have died? Bob Dylan's answer *Blowing in The Wind* was iconic of the early 1960s when I was a student. Authority was being questioned and challenged as never before. Advances were being made in free speech and civil rights. People began to think 'outside of the box' — even going beyond just being allowed to question but to having a responsibility to question. John Lennon famously said, "The thing the sixties did was to show us the possibilities and the responsibility that we all had. It wasn't the answer. It just gave us a glimpse of the possibility" (Wiener J. Lennon's Last Interview, *The Nation* (Weekly USA), 8 December 2010).

At root, I am incongruence — an academic and an engineer. Dictionaries tell us that academic means pertaining to areas of study that are not primarily vocational or applied as the humanities or pure mathematics. It implies theoretical discussion that is not practical, realistic or directly useful. In everyday usage, the word *academic* means irrelevant. Engineering is almost the direct opposite — it is practical and "down to earth". Indeed, many people think of an engineer simply as a mechanic. In this book, I want to tell you my story — my intellectual journey as an 'irrelevant mechanic'.

Like Bob Dylan and John Lennon, my intellectual awakening also began in the late 1950s and early 1960s. My perspective is, of course, rather different from that of pop stars and media celebrities. As a civil engineer, I am concerned with using theory to address very practical matters like building bridges and water supply networks. My questioning has been about the gaps I perceived between theory and practice. I have come to believe that knowing and doing have become so separated that it is harmful to our way of life. The lessons I have learned are quite general and urgent. I fervently believe that it is imperative that everyone (technical and non-technical people alike) need to do things differently if we are to address successfully some of the big challenges we face in the 21st century. My journey has taught me how practical matters are embedded in human experience. I believe that we all need to take a 'systems thinking' approach that combines traditional analytical reductionism (the idea that a whole is merely the sum of its parts) with synthetical holism (the whole is more than the sum of its parts). I maintain that this is so even when we use the most advanced of the physical sciences to meet human aspiration in engineering, medicine and all other forms of practice. I believe that systems thinking is needed if we are to create infrastructure resilient enough to cope with the possible extreme weather events of the worst predictions of climate scientists.

You could be forgiven for thinking that engineering is so highly specialized and technical that any experiences I have gained, any lessons I have learned or anything else I have to say cannot possibly be of any wider interest. To counter that, I would simply point out that engineering is at the heart of society. Engineering is very relevant to our everyday lives. Engineering is about making things. In its most general sense, engineering is about turning an idea into a reality — creating and using tools to accomplish a task or fulfil a purpose. Man's ability to make tools is remarkable. But it is his ingenious ability to make sense of the world and use his tools to make even more sense and even more ingenious tools that makes him exceptional. To paraphrase Winston Churchill "we shape our tools and thereafter they shape us". One only has to think of the railways, the telephone or the computer to realize that technical change profoundly affects social change. Engineering is a value-laden social activity — our tools have evolved with us and are totally embedded in their historical, social and cultural contexts. Our way of life, the objects we use, the understanding and knowledge we gain, go hand in hand. For example, the railways changed the places people chose to live and to take seaside holidays. Different kinds of fresh food, newspapers and mail were distributed quicker than ever before. A modern hospital is full of technology made by engineers.

Many doctors proclaim that a healthy lifestyle is easier if you have some empathy with your body and how it is performing. Likewise, you might drive better if you have some rapport with the workings of your car. A fulfilled life may be more likely if you have some harmony with the things you rely on and some feelings of why sometimes they do not perform as you might wish. So, it is in this sense that the book is aimed at general non-technical readers as well as at specialist engineers.

In summary, my purpose in writing this book is to share some of the things I have learned over a lifetime as a practitioner and an academic. I will show not only how they apply to engineering but also how they are relevant to many other aspects of life as well, including religious faith, aesthetics and fine art.

Three stages of personal life provide a theme for the whole book. They are as follows: (1) early total dependence on others such as parents and teachers; (2) later independence — the much vaunted goal of my generation to be free of the control of others — especially those on whom we have been dependent; (3) final interdependence or the often unrecognized maturity of realizing that no one 'is an island' — we are mutually reliant on others. I believe that these same stages are relevant to whole organizations, professions and complex systems like healthcare and even nation states and world government. The emphasis in the past 50 years has tended to be on issues around gaining independence. I believe that this has tended to mask the even more important and problematic move from independence to interdependence because it requires a degree of maturity built on experience. Professor Roy Severn, my boss and mentor at Bristol, expressed this eloquently to our students. He said that our job as lecturers 'was to turn them from pupils into colleagues'. By that he meant getting them not only to think for themselves (independence) but also to realize they must work with and through other people (interdependence). Unfortunately, responses to the questioning of experience, age and authority over my lifetime (especially when things have gone wrong or there has been wrongdoing) have not been so eloquent and all too often very unsatisfactory. This has led to a tendency in our western societies to value youth over experience and knowledge over wisdom. I think this is because much of the experience is tacit and the reasons for judgements and decisions are often hard to externalize. As Mark Twain wrote, “Good judgement is the result of experience and experience the result of bad judgement”.

My own intellectual progress moved slowly from total dependence on school teachers and lecturers (when I believed everything that they told me) to independent thinking as an undergraduate (largely through a lively group of fellow undergraduate students who were questioning everything). My understanding of the importance of interdependence with others came much later. Indeed, it came only after extensive experience and much reflection on becoming a parent, working as a senior manager and leader, tutoring students (as undergraduates and postgraduates) from many different parts of the world, working on engineering projects as well as the kinds of experiences we all have in our everyday lives.

An appreciation of the concept of interdependence is important, I believe, because it underpins systems thinking. As we will uncover later in the book, adopting systems thinking involves what Thomas Kuhn called a paradigm shift, i.e. a change in our basic assumptions. One of these changes is from an assumption of a kind of independence between objective facts and subjective opinion that was never made explicit in my education to what I now maintain is a natural gap of interdependence between theory and practice. I think that even the rational objective physical systems, the sciences of physics and chemistry (which many people call ‘hard’ systems, see Chapter 8), are interdependently embedded in the subjective (and often condemned as irrational) human and social systems — the ‘soft’ sciences of human behaviours. This gap of interdependence is a gap of truth-likeness, but unlike the philosophers of old who have searched for absolute certainty, I believe it to be profoundly uncertain and dependent on context. When we decide and act on our thinking our challenge is to understand and accept the nature of that uncertainty in context. We have to recognize that absolute certainty is only available through faith. The bridge over the gap between thinking and doing has to be firmly founded on an understanding that all knowledge and information is interdependent contextual evidence. This evidence can best be interpreted, modelled and processed at different levels or layers of understanding through systems thinking.

Even as a grammar schoolboy, I knew I wanted to be a civil engineer — it was buildings and bridges that fascinated me. As soon as I was old enough, and still at school, I got a summer vacation job working for a civil engineering contractor. In the following years, I worked in the office of the Derby Borough (now City) Engineer in the UK. Almost immediately, I began to wonder about the gap between what I was learning in theory at school and what was happening in practice — but of course at that time, I thought all would subsequently become clear as my studies developed. The idealized schoolboy mechanics problems of frictionless pulleys and ladders leaning against walls would evolve into real pulleys and actual ladders. I went to the University of Sheffield to read civil engineering. Then we did indeed begin to address real engineering problems. Now, I was faced with a very different set of ways to be dissatisfied about the gap. For example, we were told in our second-year structural design class that the pressure of the wind blowing on the roof of a low long building was a static pressure. This despite the fact that it is obvious to all that wind is ever changing and dynamic. So, why treat it as static? The answer was that it was simpler to calculate and that it works — many structures had been safely designed this way. Indeed, the justification for many of the ways in which structures are designed is that the methods we use work — they have resulted in safe structures in the past. I was deeply uncomfortable with this argument at the time but unable to articulate why or what we might do about it. It was later (in my early academic career) when I began to uncover my discomfort when I read Bryan Magee, summarizing Sir Karl Popper. He wrote, "Just because past futures have been like past pasts it does not follow that future futures will be like future pasts".

When in the 1970s a series of box girder bridges collapsed, I began to realize another set of reasons why Magee's summary was so important — unintended consequences. I read with particular interest the failure of the Westgate Bridge in Melbourne Australia because an undergraduate contemporary of mine was involved. I saw the effect of unintended consequences leading to loss of life. For the first time in my life (but not the last), I felt deeply that "there but for the grace of God go I". Popper was to become a very important influence on me when I discovered that he had defined social science as the study of the unintended consequences of human action. Immediately, I saw a need for a 'social science' of engineering to go alongside our well-developed 'physical science' of engineering.

My story is one of rising aspirations, awakening determination and stimulating interactions with inspiring individuals. A whole series of unintended consequences unfolded as my journey became richer and more exciting. I have chosen to tell my story by bringing together 12 of my papers, slightly modified and published between 1978 and the present. An advantage of this approach is that each chapter can be read independently but brings with it the slight disadvantage of risking some repetition — I have lightly edited the originals to reduce this. The chapters are grouped into six parts using active headings of the 'ing' form of the present participle in order to emphasize the importance of 'doing' a process. They are learning from failure, joining-up theory and practice, understanding process and classifying uncertainty, managing risks and finding resilience and finally thinking systems by integrating people, purpose and process. Each part contains a new Preamble to carry the narrative, link the chapters and identify the lessons learned.

I would like to acknowledge the many people who have influenced me and helped me directly and indirectly. First, I have been heavily influenced by writers (dead and alive) who I have never met. Principal among them are Karl Popper, Thomas Kuhn, Arthur Koestler, Bryan Magee, Peter Senge, Edward de Bono, Tom Settle and Charles Handy and it also includes many more, too numerous to mention.

Second, I thank many friends and colleagues with whom I have had many conversations, read many books and papers, discussed, corresponded and talked at length about the ideas I am putting forward in this book. The list is long and includes Bob Baird, Howard Rose, Sir Alfred Pugsley, Colin Brown, David Elms, Jim Baldwin, Bruce Pilsworth, Lotfi Zadeh, Stephan Körner Barry Turner, Nick Pidgeon, Priyan Dias, Roy Severn, Bill Smith, Norman Woodman, Jitendra Agarwal, Colin Taylor, Patrick Godfrey, Jim Hall, John Davis, Oksana Kasyutich, Paul Jowitt, Allin Cornell, Ted Galambos, Bruce Ellingwood, Ross Corotis, Rob Melchers, John Caldwell, Arturo Bignoli, Alberto Bernardini, Weicheng Cui, Mauricio Sanchez-Silva, Lorenzo van-Wijk, Dick Taylor, Stewart Craddy, Simon Pitchers, Vaughan Pomeroy, Mike Barnes, Keith Eaton, and Michael Beer.

Thank you to Lance Sucharov, Jennifer Brough, and Aanand Jayaraman at WSPC for turning my efforts into the book you see.

Last but by no means least, I thank my wife Karen for her love and support — without her, none of this would have been possible.

Books for the General Reader

1. *Engineering: A Very Short Introduction*, 2012, Oxford University Press, UK.
2. *Structural Engineering: A Very Short Introduction*, 2014, Oxford University Press, UK.

About the Author

David Ian Blockley is an Emeritus Professor of Engineering at the University of Bristol, UK. He was born in Derbyshire, England, in 1941. He graduated from the University of Sheffield in 1964 and gained his PhD there in 1967. For 2 years, he was a Development Engineer with the British Constructional Steelwork Association in London working on design development. In 1969, he became a Lecturer in Civil Engineering at the University of Bristol, was promoted to a readership in 1982 and a personal Chair in 1989. In the same year, he was appointed Head of the Department of Civil Engineering, University of Bristol until 1994 when he was elected Dean of Engineering at the University of Bristol from 1994 to 1998. He was reappointed as the Head of Civil Engineering in 2002 until his retirement in 2006.

He is a Fellow of the Royal Academy of Engineering, of the Institution of Civil Engineers, of the Institution of Structural Engineers and of the Royal Society of Arts. He was the President of the Institution of Structural Engineers during 2001–2002. He is a Corresponding Member of the Argentinean Academy of Science and of Engineering. He was a Non-Executive Director of Bristol Water plc from 1998 to 2007. He is the author of over 180 papers and seven books, and has won several technical awards for his work, including The Telford Gold Medal of the Institution of Civil Engineers. His first four books were technical and written for other engineers, but his most recent three books have been written for the general reader, namely *Bridges*: *The Science and Art of the World's Most Inspiring Structures* (2010, Oxford University Press, UK), *Engineering*: *A Very Short Introduction* (2012, Oxford University Press, UK) and *Structural Engineering*: *A Very Short Introduction* (2014, Oxford University Press, UK).

Contents

The 20 Learning Points

No.	Learning Points
1	Failure is an opportunity to learn.
2	Engineers require safe theories whereas scientists need to prove them wrong.
3	We should consciously try to minimize any unintended consequences of a decision.
4	We should replace the reliability of an idea with a responsibility to act on that idea.
5	Academic institutions are currently dominated by technical rationality.
6	Innovation is key to future success — we need to think outside of the box.
7	Change is a 'new' process of becoming different.
8	Purposeful new process is the key to integrating people, purpose and 'old' process.
9	Systems thinkers build models based on three basic concepts: (1) layers of (2) connected and (3) new processes.
10	Interactions between processes create emergent properties or characteristics.

No.	Learning Points
11	Soft systems are difficult to predict — use dependable evidence to achieve purpose.
12	The three orthogonal structural attributes of uncertainty are fuzziness, incompleteness and randomness.
13	We need to distinguish how the courts allocate blame from the lessons learned.
14	The Italian flag can help judgements about incomplete knowledge.
15	Complex systems may contain new risks through unknown interdependencies.
16	Knowing and doing have been systematically separated — we need to bring them together to create resilience.
17	Integrate reductionism and holism to create synergy.
18	Delivering resilience will require systems-thinking skills that go beyond technique.
19	Bridges are built by people for people.
20	The book of a bridge is analogous to the book of a piece of art or an organism.

Part I

Learning from Failure

Preamble

No one likes to fail — few of us see failure as a positive. Yet, from the moment we are born, things do not always work out as we want. Most of us soon learn to live with that as part of the ups and downs of everyday life. It is a commonplace truism that we should learn from our mistakes. Yet, to do that, the first requirement is perhaps the hardest — admitting them. Many of us spend a lot of time in denial. However, I think it is how we respond to mistakes and failure that is critical. Real failure is not in the 'falling-down' but in the 'staying-down'. Resilience or the ability to 'get-up-again' and bounce back is the key to learning from our mistakes and failures.

Unfortunately, we live in a society where achieving success seems to be everything and failure is a calamity. This tends to reinforce our natural and long-standing craving for certainty. Since the ancient Greeks, collectively we have wanted to know the 'objective' truth about everything. Some of us find our certainty through religious faith, some through science and the rest of us deal with our uncertainties as best we can. Since the time of the ancient Greeks, the relationship between theory and practice has been controversial. We have sought theory which is value-free and with a perfection sometimes so removed from reality that it tells us little about how we should live our lives.

It is my contention that, over the centuries, our western culture has systematically tended to separate theory from practice in a manner that is actually harmful to our way of life. We have been set up to fall harder when we do fail. However, it is a myth that successful people never fail. Being negative about failure can prevent us from gaining the resilience we need. Successful people are usually resilient — they learn from their mistakes and failures and they learn to cope with uncertainty. Indeed, they see opportunities in uncertainty and regard risks and opportunities as the opposite sides of the same coin. I maintain that these are characteristics that we will all need as we face the possibility of more extreme weather events if the predictions of the climate scientists are anywhere near the mark.

Public and professional practice is quite different because all decisions have to be justified — often in public. Professionals can and will be challenged, quite rightly, and especially if something goes wrong. As a consequence they must use the latest and most dependable evidence, including science, to inform their decisions.

As I said in the Preface, the three stages of personal life (dependence, independence and interdependence) provide a theme for the whole book. My own intellectual progress from total dependence on teachers to independent thinking moved slowly. I found the school GCE 'A' level approach to physics and applied mathematics very unsatisfactory. I was not motivated by science. Frictionless pulleys and other idealistic assumptions taken to make the problems tractable left me cold. They were, of course, left behind in my undergraduate engineering course. Unfortunately, others took over. The 'rules of thumb', 'fudge factors' and other empiricisms we were told to use to cover the uncomfortable differences between the theory and the reality were only a marginal improvement. They were simply justified because they worked. We were told that this was how it was done — the result of experience of trial and error. Of course, there was an underlying unarticulated assumption — because they had worked in the past, then they would work in the future. There was no questioning of the contexts within which they were known to work had changed. The rules were simply the best we can do.

In the mean time, I had become acutely aware that, after graduation, when we get out into the real world of engineering design, our calculations would *actually* be tested by nature. This was not pretence any more — it was real. So, if you got the calculations for a beam in a building wrong, then it might well collapse and kill someone. In that new reality, I felt a burden of responsibility I had never faced before. But again, no one mentioned what I later found to be rather important — the notion that all professionals cannot always be right but they have a legal duty of care — a duty not to act negligently — an obligation to take all reasonable care to avoid harm to others.

When I joined the staff of the Civil Engineering Department at the University of Bristol in 1969 I began to read about structural failures. At that time, in the early 1970s, they were making some headlines. In particular, a number of box girder bridges failed (see Chapter 1).

I began to seriously question and think about the gaps between what engineers know, what they do and why things go wrong. I began to see that our calculations were not really scientific. Rather, they were based on science but embedded in so much uncertainty that I coined the term 'calculational procedural model' to capture that idea. Up to that time, I had only heard the word 'model' used only about toy cars and physical specimens — it was much later when I realized its use in describing a theory is of crucial importance. I also became aware of a paper written by Sir Alfred Pugsley in 1969 called "The Engineering Climatology of Structural Accidents". It was the first International Conference on Structural Safety and Reliability (ICOSSAR) held at Washington in 1969. Sir Alfred was an Emeritus Professor of my department and so I had several opportunities to talk to him. He opened my eyes to some of the wider issues (that he called the climate), such as excessive pressures on the project from international, national or local, financial, political or industrial conditions. He was reticent, very formal and private man but very perceptive and insightful. His paper gave me the confidence to think and write openly about issues which had previously seemed contextual but which I began to see were in fact part of the system we should be examining and controlling. I was later to be part of the organizing scientific committee for several ICOSSARs for my technical work though unfortunately few other researchers took up Pugsley's lead.

In the first three chapters that form Part I, my purpose is to show how success and failure are interdependent right 'across the board' from individuals to large-scale infrastructure projects. **Learning Point No. 1** in the first chapter came from Pugsley's work — 'failure is an opportunity to learn'. Microsoft's Bill Gates knows this — he is reported as saying, "Your most unhappy customers are your greatest source of learning".[a]

Chapter 1 was first published in 1977, yet the 'message' is still relevant in 2019. The chapter was a 'game changer' for me. I learned a little later (Chapter 2) that the philosopher/scientist Thomas Kuhn used the term 'paradigm shift' to capture the more general idea of a game change in worldview or set of practices that define a discipline. Previously, in the whole of my educational, industrial and academic experience, no mention was made of the need to address people issues. Through the work I did for

---

[a] Bill Gates quoted at https://www.brainyquote.com/quotes/bill_gates_161558.

the paper, I began to understand the importance of human factors in determining risks. Of course, common sense tells you that you must work with and through others. But previously, there was an implied unstated set of assumptions that engineering is an objective rational discipline based on firmly founded physical science with little doubt. Judgements were subjective and often irrational and unreliable. Yet, I came to realize that practitioners were constantly making them because they had to.

Three changes to the way I worked followed the publication of the paper. Firstly, through it, I was awarded The Telford Gold Medal of the Institution of Civil Engineers in 1978 for the best paper published in their various journals. This meant that people noticed what I was doing and I was having an influence. That encouraged me to the second change which was to include a project for all second-year civil engineering undergraduates at Bristol in which they chose a famous failure (like Ronan Point [1]), studied it, gave a talk to the class about it and wrote a report which I assessed for coursework. I gave the students the list of factors that appear in the paper and of course they had access to the paper. The project amended over the years is, I am told, still set today. The third change was to my emerging research agenda. I knew I needed to work with social scientists (Chapter 3) to get any further on human factors, but I also had to get a better philosophical (Chapters 2, 4, 5, 7, and 8) and mathematical (Chapters 8–10) grasp on uncertainty.

History does have a habit of repeating itself. Stories of structural failures since 1977 could be included in the original study without changing the conclusions. For some examples, read the stories of the Citycorp Centre in New York (1978),[b] the Hyatt Regency Walkway (1981),[c] Chernobyl (1986),[d] the London Millennium Bridge (2000) [2], the I-35 Highway Bridge Minneapolis (2007)[e] and the Collapse of the Roof of a Supermarket in Riga, Latvia (2013). There are many more such as the Genoa Bridge (2018) [3] and the fire at Grenfell Towers (2017) [4].

---

[b] See http://www.onlineethics.org/cms/8888.aspx.

[c] See https://nvlpubs.nist.gov/nistpubs/Legacy/BSS/nbsbuildingscience143.pdf.

[d] See http://www.world-nuclear.org/info/Safety-and-Security/Safety-of-Plants/Chernobyl-Accident/.

[e] See http://en.wikipedia.org/wiki/Riga_supermarket_roof_collapse.

The paper demonstrated in 1977 how we can learn from studying case histories of failure of major infrastructure projects to develop different ways of assessing degrees of proneness of a structure to an accident. But one paper or even the number of books and papers published before and since 1977 have not proved to be sufficient grounds to mount legal proceedings against those culpable. A more important goal is to try to ensure the lessons are absorbed and applied in engineering practice. Many other university courses now also require undergraduate students to study and learn from case histories of engineering failures.

In 1978, Richard Henderson was the first person willing to trust me as his supervisor for his PhD. For that, I have always been grateful to him. Working with Richard, I learned a great deal about project supervision and together we wrote Chapter 2 published in 1980. In 1981, we were awarded The George Stephenson Medal of the Institution of Civil Engineers. **Learning Point No. 2** in Chapter 2 is about the essential differences between engineers and scientists. 'Engineers require safe theories whereas scientists need to prove them wrong'. Again, I believe that Chapter 2 is timeless and applies now just as much as it did then. The lessons follow directly from those of Chapter 1. Richard and I built on the philosophical writings of Karl Popper and Thomas Kuhn.

We demonstrated that failure is essential for progress and the growth of knowledge. We explored the implications for the differences between progress in controlled scientific testing and in engineering where failure is to be avoided because it is disaster. We describe how the nature of the interdependence between success and failure is logically asymmetrical. We show that when we fail we know for sure that our reasoning was wrong or, as the case history in Chapter 3 demonstrates, we failed to prevent some unintended consequences. But when we succeed then, logically, our reasoning could have been right or wrong. If we were right then fine but if we were wrong then we were just lucky or other unknown factors came into play. In essence, the asymmetry makes getting beyond the extreme sceptic (an evil demon who doubts everything) to an absolute truth that applies in all circumstances and contexts very difficult. Therein lies some deeper learning. Truth we will find depends on context and understanding that is crucial to understanding the differences between the scientific search for truth and managing the risks of engineering practice.

We argued that opportunities of learning from failure are crucial, that engineering theories are only 'weakly falsified' and that legal processes often inhibit the potential to learn. We concluded that the development of engineering rules is therefore much slower than the development of scientific knowledge and is part of the set of reasons why some scientists condemn engineering science as being *ad hoc* and lacking in rigour. These thoughts were the foundation of my much later thoughts about the nature of practical rigour (Chapters 6 and 12).

It was whilst doing this work that I realized the need to work with social scientists to try to develop a 'social science' of engineering. I first met Barry Turner in the early 1980s — he was, at that time, a Reader in Sociology at the University of Exeter, UK. As a student he had done one undergraduate year of an engineering course before changing to sociology and so had some empathy with my goal. We were able to find some funding from the foresightful but short-lived initiative of a Joint Committee of the UK Economic & Social Science Research Council and Science & Engineering Research Council. Barry and I employed Nick Pidgeon, a psychologist who had just completed his PhD, as a research assistant. Barry, Nick and I worked together until Barry's untimely death in 1995. In 2018, Nick is Professor of Environmental Psychology and Director of the Understanding Risk Research Group at the University of Wales in Cardiff. Chapter 3 is one of the many papers we wrote together. It was one of the first ever case histories ever conducted by engineers and social scientists working together. It is the story of an actual failure of a factory building which I helped consulting structural engineer Stewart Craddy to investigate and advise the owner of the building. At first, I thought the technical reasons for the failure were so obvious that the cause of the collapsed roof was almost trivial. However, working with Barry and Nick helped me to see the context was not trivial at all and led to **Learning Point No. 3** that 'we should consciously try to minimize any unintended consequences of a decision'.

The story illustrates that engineers can so easily misjudge the effects of changing trends in practice. For sound commercial reasons safe loads for roof purlins were increasingly being based on ultimate load tests using code-specified loadings. At the same time, the effects of a step in the roof line or the use of a parapet wall meant a greater likelihood of snow drifts that previously had not been considered. Minimum acceptable standards written into codes almost certainly were becoming the normal standards through economic pressures. All of these changes were, to use the term coined by Barry Turner, 'incubating' the preconditions to failure. As I explain in Chapter 11 imagine the development of a failure as analogous to the inflation of a balloon. The start of the incubation is when air is first blown into the balloon, when the first preconditions for the accident are established. The pressure of air in the balloon represents the 'proneness to failure' of the process. Events accumulate to increase the predisposition to failure. The size of the balloon can be reduced by letting air out and this parallels the effects of management decisions that remove some of the predisposing events and reduce the proneness to failure. If the pressure of events builds up until the balloon is very stretched then only a small trigger event, such as a pin or lighted match, is needed to release the energy pent up in the system. The trigger is often confused with the cause of the accident. The trigger is not the cause — the overstretched balloon represents an accident waiting to happen. In accident prevention, it is important to recognize the preconditions — to recognize the development of the pressure in the balloon.

The obvious next question is how can we measure these incubating preconditions — these 'balloon' pressures? Chapter 1 was written before I met Barry Turner, so the measures I used on the case histories were simple point scores following along similar lines to Pugsley's seminal paper on the climatology of structural accidents. However, I knew point scores were inadequate because of the complex interdependencies between the parameters. There was research to be done to find a metric. In the mid-1970s, Jim Baldwin of the Engineering Mathematics Department in Bristol gave me a copy of a paper by Lotfi Zadeh on a new idea called a fuzzy set. I was immediately intrigued because it was an intuitively attractive idea and I saw it had great possibilities. Zadeh was an Iranian American control engineer based in Berkeley. Zadeh wrote '… as the complexity of a system increases, our ability to make precise yet significant statements about its behaviour diminishes until a threshold is reached beyond which precision and significance become almost exclusive characteristics'. His ideas took root very quickly with whole journals and conferences devoted to this new approach. In 1975, I published a paper in which I was the first to suggest a way in which fuzzy sets could be used in a Civil Engineering problem. As a result, I had invitations to visit many Universities and speak at a number of conferences around the world.

Now, in Chapter 1, we examine some case histories of structural failure.

References

[1] Ministry of Housing and Local Government (1968). *Collapse of Flats at Ronan Point, Canning Town*. HMSO, London.

[2] Blockley, D. I. (2010). *Bridges*. Oxford University Press, Oxford.

[3] Genoa Bridge Collapse (2018). *Ponte Morandi: Report of the Inspection Commission Mit.* Ministero delle Infrastrutture e dei Trasporti, Commissione Ispettiva Ministeriale, Italy. See http://www.mit.gov.it/comunicazione/news/ponte-crollo-ponte-morandi-commissione-ispettiva-genova/ponte-morandi-online-la (Accessed on 31 May 2019).

[4] Grenfell Towers (2018). *UK Government Collection Grenfell Towers.* See https://www.gov.uk/government/collections/grenfell-tower (Accessed on 31 May 2019).

Chapter 1

Analysis of Structural Failures*

Abstract

I present a classification of basic types of structural failure and then expand it into a set of statements which could be assessed subjectively as the parameters of a prediction process. This process is intended to account for a structure failing due to causes other than stochastic variations in load and strength. The parameters are assessed for 23 major structural accidents and one existing structure, and are analyzed using a simple numerical interpretation. The accidents are ranked in their order of inevitability. Human errors of one form or another proved to be the dominant reasons for the failures considered.

Introduction

There has been in recent years an upsurge of interest amongst engineers in matters related to structural accidents. Reports of inquiries into recent accidents have become compulsive reading, whilst at the same time, the redrafting of codes of practice into the limit state format has stimulated inquiry into the use of probability theory to determine suitable partial factors. An increasing concern about the way actual structures behave rather than

*This chapter was originally published in *Proc. Instn. Civ. Engrs., Part 1*, 1977, 62, 51–74 and presented at an ordinary meeting, 5.30 p.m., 15 March 1977. Written discussion in *Proc. Instn. Civ. Engrs., Part I*, 1977, 62, 681–736.

idealized theoretical models or isolated laboratory tests on physical models or elements of structures is another aspect of this interest.

One of the major difficulties in discussing this subject is the problem which has existed since Galileo, the communications gap that often exists between the practising engineer and the research worker. Many engineers believe that, for instance, probability theory will be of little help in understanding the basic causes of structural failures. Slightly altering the values used for safety factors (be they total or partial) will be of little consequence in this regard. It is difficult, on the other hand, to wrap up problems such as poor site control, errors of judgement and pressures due to shortage of time and money, into mathematical models or rules of procedure to be written into codes of practice.

Probability theory is a mathematics of uncertainty. Its use requires a different way of thinking about mathematics than the traditional mathematics most engineers were taught. Many engineers cannot see any reason why they should bother. Most research workers in the field of structural safety would agree that probability theory is an essential tool in helping to predict the actual behaviour of a structure.

The purpose of this chapter is to review and classify some structural failures and to consider ways in which predictions of the likelihood of possible future structural accidents can be performed.

Design Process

In order to discuss a classification of structural accidents, it may be instructive first briefly to examine the design process in a very general overall way. Figure 1.1 shows a simplified decision tree demonstrating the overall decision routes facing the structural designer. This is intended to be a conceptual model made at the design stage of the general decisions yet to be faced. The first two decision paths to be taken (those of alternative structural forms) are the designer's choice using mainly professional information. The third path, the construction process, will be partially under the control of the contractor and designer, and the final path, the use of the structure, will be largely the consequence of these earlier decisions.

There are obviously many overall design solutions to any problem. A single-storey shed may, for instance, be a concrete portal, a steel portal or a steel truss: this is the first decision path. There is a multitude of detailed

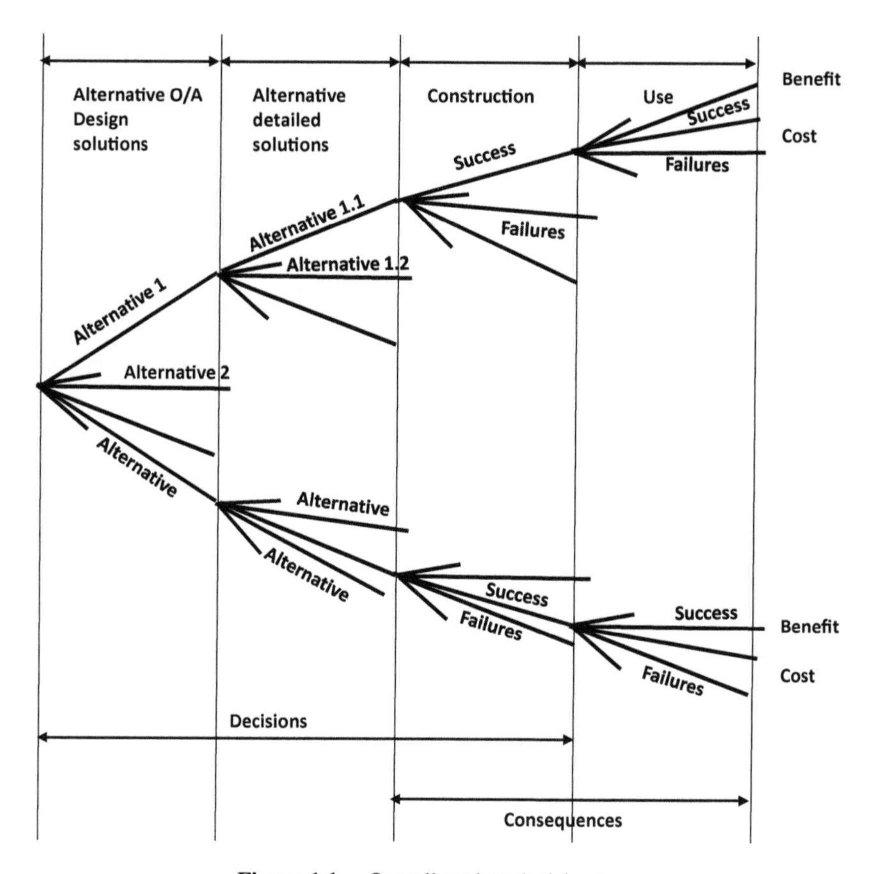

Figure 1.1. Overall project decision tree.

design decisions concerned with the spacing of frames, the lateral bracing, design of joints, etc., which are represented by a single line in Figure 1.1: the second decision path. If any one of these paths is taken, costs will be incurred. It is the designer's task to minimize these costs. Probability theory can help with these decisions [1]. The designer has to make these decisions so that it is a virtual certainty that the structure will be constructed successfully and will be used successfully. However, there is always a small, but finite, chance or probability that one of the events listed in Table 1.1 will occur and the structure will be damaged or collapse, and a cost penalty incurred.

The central problem of the designer is to design his structure economically and aesthetically so that the probability of each of these unacceptable events

Table 1.1. Some causes of structural failure.

Limit states	**Overload:** geophysical, dead, wind, earthquake, etc.; man-made, imposed, etc. **Understrength:** structure, materials, instability **Movement:** foundation settlement, creep, shrinkage, etc. **Deterioration:** cracking, fatigue, corrosion, erosion, etc.
Random hazards	**Fire, floods, explosions:** accidental, sabotage earthquake, vehicle impact
Human-based errors	**Design error:** mistake, misunderstanding of structure behaviour **Construction error:** mistake, bad practice, poor communications

or accidents is acceptably low. In order to make his decisions, the designer has certain tools at his disposal. These tools are basically his knowledge of structural theory, the research and development information available and his professional knowledge and experience. However, he is also subject to constraints such as government regulations and client budgets. Having chosen the overall type of structure using his professional knowledge, the designer has to make his detailed design decisions on the basis of a theoretical model coupled with research and development information, codes of practice, etc. The classical way of avoiding the unpleasant events labelled as limit states in Table 1.1 is to use the well-known working stress concept. Ultimate load and now limit state concepts are more modern approaches. Here the decisions are based upon a calculation procedural model (cpm) which is usually the combination of elastic theory or plastic theory, with research and development information often distilled by research workers and various committees into recommended procedures.

Certain of these calculation procedural models are good in that they are well proven by analysis and laboratory testing, usually on elements of the final structure, and designers can use these procedures with some confidence. Much of the difference between the performance of the structural element in the laboratory and in the structure is due to the random nature of the parameters describing the system. Here then the probability of an extremely high value of load or an extremely low value of strength causing collapse can be calculated with some confidence if the variabilities of the various parameters are known. This has been done, for example, on steel portal frames and concrete slabs [1, 2].

In contrast, the behaviour of a foundation on a compressible soil or the cumulative damage of fatigue loading cannot be predicted at all accurately. In this situation there are two types of uncertainty: the system itself is poorly understood; and there is a high random variability of the parameters describing the system. In trying to understand the behaviour of his sort of system probability theory could well be used much more in the basic research work.

However, designing to the limit states of Table 1.1 is only part of the problem. The precautions to be taken by the designer against such random hazards as fire and earthquakes obviously depend on the likelihood of occurrence of the hazard and its consequences. Thus, fire is a major hazard to all buildings, but earthquakes in Britain for most structures can be neglected. However, the consequences of failure of a nuclear reactor due to an earthquake, could be so enormous that the possibility of earthquake loading has to be considered.

The designer tries to ensure that the third group of events in Table 1.1, the human-based errors, does not occur. He does this by good professional practice and communication. He does not take the possibility of a failure due to this sort of reason directly into account in his design calculations. However, in solving a particular problem, he may recognize that certain structural solutions may be more susceptible to this type of error than others although this is a factor difficult to quantify and would rarely change a major design decision.

Under normal circumstances, the designer cannot make his design decisions with detailed knowledge of the contractors' plans, but of course the good designer is always aware of possible construction problems. Human error during the construction phase can, for the most part, be prevented by good communications between all parties concerned and well-defined responsibilities under the contract.

Classification of Failures

With this discussion in mind and based on a reading of reports of 23 structural failures [3–27] and more general references [28–36], a classification of basic types of structural failure was formulated. Clearly, any one accident will in general involve a number of basic types. Errors may occur at any stage in the decision route (Figure 1.1) and will generally result in one of the undesirable events listed in Table 1.1.

The basic types proposed are as follows:

(a) structures, the behaviour of which are reasonably well understood by the designers (and consequently the calculation procedural models are good), but which fail because a random extremely high value of load or extremely low value of strength occurs (e.g. excessive wind load, imposed load, inadequate beam strength);
(b) structures which fail due to being overloaded or to being under strength (as 1), but where the behaviour of the structure is poorly understood by the designer and the system errors in the calculation procedural models are as large as the random errors in the parameters describing the model; the designer here is aware of the difficulties (e.g. foundation movements, creep, shrinkage, cracking, cumulative damage, durability generally);
(c) structural failures where some independent random hazard is the cause, e.g. earthquake, fire, floods, explosion, vehicle impact; the incidence of this type can be obtained statistically;
(d) failures which occur because the designers do not allow for some basic mode of behaviour inadequately understood by existing technology (this mode of behaviour has probably never before been critical with the type of structure under consideration; a basic structural parameter may have been changed so much from previous applications that the new behaviour becomes critical, or alternatively, the structure may be entirely of a new type or involve new materials or techniques; it is possible, however, that some information concerning the problem may be available from other disciplines or from specialist researchers, and this will be information which has not generally been absorbed by the profession);
(e) failures which occur because the designer fails to allow for some basic mode of behaviour well understood by existing technology;
(f) failures which occur through an error during construction; these would be the result of poor site control, poor inspection procedures, poor site management, poor communications leading to errors of judgement, the wrong people taking decisions without adequate consultation, etc., and may also occur through a lack of appreciation of critical factors and particularly through poor communications between designers and constructors;

(g) failures which occur in a deteriorating climate surrounding the whole project; this climate is defined by a series of circumstances and pressures on the personnel involved; pressures may be of a financial, political or industrial nature, and may lead directly to a shortage of time or money with the consequent increased likelihood of errors during both design and construction processes; they may also result in rapidly deteriorating relationships between those involved in the project;
(h) failures which occur because of a misuse of a structure or because the owners of the structure have not realized the critical nature of certain factors during the use of a structure; associated failures are those where alterations to the structure are improperly done.

Although it would be interesting to classify all structural failures into these basic types, the information gained would not be directly useful. A more detailed examination of each type is required and statements which will be useful not only in studying past failures but also in predicting the likelihood of future failures need to be formulated. This checklist of failure parameter statements, now presented, is based on the basic failure types proposed and may be used in a prediction process [37]. A total of 23 past structural failures will be assessed, for calibration purposes, using these parameters. These assessments will also enable certain conclusions to be drawn regarding the major reasons for past failure. It will be appreciated that these assessments can only be made subjectively and thus, for this exercise, the parameters were formulated as statements and the assessor (the author in all cases but one) was asked to assess his degree of confidence in the truth of and the importance of the statements.

A Checklist for Assessment of Likelihood of Structural Accident

A checklist of statements has been devised, and the procedure is as follows. Assess (a) the degree of confidence in the truth of and (b) the importance of the following statements. Choose any two of the following descriptions:

1 — very high confidence	A — very low importance
2 — high confidence	B — low importance
3 — medium confidence	C — medium importance
4 — low confidence	D — high importance
5 — very low confidence	E — very high importance

1a. The loads assumed in the design calculations are a good and/or a safe representation of the loads the structure will actually experience.
1b. Any variabilities in the values assumed for the parameters used to describe the strength of the structure have been well catered for.
2a. Assuming that the design calculations have covered all possible failure modes for the structure, the calculation procedural model is a good representation of the way the structure will behave if constructed to plan.
2b. The quantity and quality of research and development available to the designer is sufficient.
3a. The information available regarding the likelihood of such random hazards as earthquakes, fire, floods, explosions, vehicle impact is sufficient.
3b. The structure is not sensitive to the phenomena of 3a.
4a. The materials to be used in the structure are well tried and tested by use in previous structures.
4b. There are no possible effects which could occur in the material which have not been adequately catered for in the design calculations.
4c. The form of structure has been well tried and tested by its use in previous structures.
4d. There is no step change in the values of the basic parameters describing the structural form from those values adopted in previous structures.
4e. There is no possible danger of a mode of behaviour of the structure, inadequately understood through existing technology and which has never before been critical with this structural form, now becoming critical.
4f. There is no information concerning the materials or the structure which is available in other disciplines and which could have been used in this design calculation.

5a. There are no errors in the calculation procedural model or possible modes of behaviour which are well known through existing technology but which have been missed by the designers.
5b. Assuming the design is based upon a good calculation procedural model the likelihood of calculation errors is negligible.
5c. The designers are adequately experienced in this type of work.
5d. The personnel available for site supervision are adequately experienced.
5e. The design specifications are good.
6a. The construction methods to be used are well tried and tested (including off-site fabrication).
6b. The structure is not sensitive to erection procedures.
6c. The likelihood of construction errors is negligible.
6d. The contractor is adequately experienced in this type of work.
6e. The contractor has personnel available for site work and supervision capable of appreciating the detailed technical problems associated with the design.
7a. The contractual arrangements are perfectly normal.
7b. The general climate surrounding the project design and construction is perfect under each of the following headings: financial, industrial, political, professional.
8. The structure is not sensitive to the way it is used.

In parameter 1a, it is presumed that the designer always tries to choose a representation of the actual loads on his structure which is conservative and safe. A critical form of loading is more likely to be missed if the representation of the actual loads is based on a poor model. For example, the use of equivalent static loads for dynamic loading situations could easily lead to trouble where unusual circumstances produced resonance or large dynamic magnification. In parameter 1b, the aim is to cover adequately statistical variations in strength values which comply with the specifications and which are catered for generally by the use of appropriate safety factors. Parameters 2a and 2b are included to question the degree of confidence of the assessor in the model used for the design decisions, that is effectively the degree of confidence in the use of elastic and/or plastic theory, research and development information and codes of practice. This can be contrasted with 4e which deals with the possibility of a completely new form of structural behaviour taking over, such as occurred with the famous Tacoma Narrows Bridge [7]. Parameter 5a deals with the calculation procedural model but concentrates on the possibility of human error. Parameters 5b–5e are also aimed at the possibility of human error. Parameter 5b asks effectively if the designers are working in good conditions, with good communications between the various members of the design team and good management procedures. Are the calculations numerically complicated and how reliable are the computer programs used? Parameters 5b, 5d and 5e are intended to cover design office to resident engineer (if any) communication and communications between resident engineers and contractor, both on the site and through the specification. Relevant subsidiary questions to 6c would be, for example, does the contractor have a good safety record, good management procedures, good labour relations with no evidence of slack site control? Has a low tender bid produced excessive pressures (financial or shortage of time)? Parameter 7a is intended to cover the contractual arrangements and whether the responsibilities of the various parties to the agreement have been adequately defined. Pugsley [34] has discussed those factors covered by parameter 7b under the heading 'engineering climate'. This parameter is intended to check whether there are excessive pressures on the project from international, national or local, financial, political or industrial conditions. Parameter 8 is included because for some special structures the user may need some educating with regard to critical factors in the structure.

Analysis of 23 Structural Failures

The parameters outlined above were formulated to assess the likelihood of a structural accident in the future. In this section, the values of assessments made by the author on 23 failures [3–27] are presented (Table 1.2). Assessments for parameter 4f were not made. The analysis was purposely limited to a number of accidents where detailed inquiries into the nature of the failure were reported. Thus, reports in engineering news journals were not considered. It is appreciated that this selection of accidents is not a random sample. The assessments are entirely the subjective opinion of the author, based upon a reading of the references. It is therefore quite possible that engineers more qualified to consider any particular failure would make slightly different assessments. However, it is felt that general conclusions may be drawn from the results presented.

Certain difficulties arise in assessing statements formulated for future projects when considering events of the past. This is particularly highlighted in parameters 5d and 6e when trying to assess the competence of the personnel running the site and inspecting the works. In assessing a future project one can only make a judgement on the basis of past performance; when assessing a failure one can assess actual performance where this is discussed in the report. In some accidents, e.g. Kings Bridge, the designers were not represented during fabrication or construction under parameter 5d and inspection was done by the client. The assessment made here is based upon the inspection done on behalf of or by the client. Parameters 5d, 6c and 6e are often difficult to separate (as for the accidents at Listowel and Aldershot), but it is assumed from the way the structure failed that all three aspects were deficient. In several of the accidents the design work was done by the contractor either as part of a package deal (e.g. first Quebec Bridge) or because the failure was part of some supporting falsework (e.g. Lodden Bridge). The assessments for these failures were based on the performance of the designers who designed that part of the structure which failed.

The failure of the Point Pleasant Bridge [9] indicates the difficulty of preventing some failures. Built to specification, this American suspension bridge was completed in 1928 and failed in 1967. The cause of the failure was a fracture in an eye bar link resulting from a crack which had own through stress corrosion and corrosion fatigue. The crack was impossible to detect without disassembly of the joint.

The failure of the Bedford Town Hall [18] is an unusual example of a design failure. The design calculations were hopelessly incorrect and were performed by an engineer working in isolation. Any one of a number of mistakes included in parameter 5a could have caused failure. In most other accidents the mistakes were much more difficult to isolate. Thus, there was a high concentration of errors in one parameter.

It is impossible, due to limitation of space, to discuss fully the reasons for each of the assessments in Table 1.2. Only a brief mention can be made, as an illustration, of the assessments for the Quebec Bridge.

Table 1.2. Accident parameter assessments.

	Tay	Quebec 1	Quebec 2	Tacoma	Kings Bridge	Point Pleasant	Westgate	Second Narrows	Heron Road	Lodden	Aroyo Seco	Listowel
1a	5E	4D		3D								5E
1b	3E		5E									
2a	4E	5E		4E		2D	5D		3B	3B	3B	
2b	4C	5E		5E	3C	5E	5C		3D	3D	3B	
3a												
3b												
4a	2B				5E	4E						
4b	4E			2A	5E	5E						2A
4c							2C			2E		
4d	4C	5E		5E								
4e	2C	5E		5E			5E					
5a	5E	4E			3E		4D	5E	5E	4E	5E	
5b		3D				2A	5E					4C
5c	2B	3C			2B			4E				2B
5d	5E	4E			5E		5E	3D	3E	5E	4E	5E
5e	4E	4E			3E		4C					5A
6a	3A	2D					5E					
6b	5A	4E	4D				5E					
6c	5E	4D			5E		4C			5E	5E	5E
6d	3E	2B			2D		3B					4C
6e	5E	4E			5E		5E		3E	5E	5E	5E
7a	2C	4E			4E		4C					4C
7b	5E	3C			3C		5E					4B
8	5E										2B	4B

	Aldershot	Bedford	Ronan Point	Camden	Stepney	Ilford	Ferrybridge	Mt Gambier	Sea Gem	Ardeer	Transocean III	Example
1a			5D				5E		4C		4C	1E
1b												1E
2a	4E		5E	5E			4D	3E	4C	4E	4D	3D
2b	3C		4D	3C	4D	4B	4D	4D	3C	4D	3C	3D
3a			3C									2B
3b			5E									2B
4a				5D	5E							
4b				4D	5E	4C		5E	5E	5E	4E	
4c	3A		3C						4D		4E	
4d	5E		3D				3C					
4e						4D	5E			5E	3B	
5a	5E	5E	5E	5E	5E	5E		3E	5D		5E	
5b		5D										
5c			2B									
5d	5E		3D		4E	5D			5E			
5e	5C						5E					
6a					3E							
6b	2C							5E				2B
6c	5E		3C	2C	4E	5E			5E	5E		2C
6d												
6e	5E		3C		4E	5D			5E			
7a	4B		4D				3E			3C		
7b			4D							2B		2A
8					4C	4C			4E	3C	5E	

Quebec Bridge

The Quebec Bridge was a steel cantilever structure over the St. Lawrence River which failed during construction in 1907. A second bridge suffered the failure of a suspended span in 1916, but was finally completed in 1918.

An error in the calculation of the dead load on the first Quebec Bridge resulted in dead load stresses of the order of 7% in excess of those calculated. This error has been included in the assessment of parameter 1a as it concerns loading rather than structural response. The state of knowledge around the turn of the century with regard to the behaviour of latticed columns was very uncertain and the research information available to the designers was very limited. No proof testing was done on the made-up column sections for the first bridge and higher permissible stresses than were usual in bridge construction were adopted. The importance of the end details and splices and the effect of lack of straightness in the columns were not fully appreciated. Consequently, the assessments made for parameters 2a, 2b, 4e and 5a were all serious. The chord which failed in the Quebec Bridge, when compared to five other American cantilever bridges of the time, had considerably less horizontal stiffness (*l/r*), less lattice area, less rivet area and less splice plate area in proportion to the size of the members and the assessment of parameter 4d reflects this.

Parameters 5b–5d are concerned with the other conditions surrounding the design of the bridge. Besides the dead load calculation error other minor calculation errors occurred. Theodore Cooper, the highly respected but ageing consultant, was not responsible for the actual design (which was in fact done by the contractors), but because he checked all the calculations a feeling of false security amongst the design staff was engendered. The actual designers' experience was based on small bridge practice. The site supervision was inadequate and there was a confusion of responsibility between the chief engineer of the Quebec Bridge Company and Cooper, the consultant. Cooper rarely visited the site, but advised frequently. The specifications were written by an inexperienced engineer and there was nothing to indicate that the bridge was an exceptional structure. The specifications were in fact based on small bridge design with tolerances which were too tight.

Parameters 6a–6e deal with the construction of the bridge. As this was a cantilever structure it was obviously at a very sensitive stage during erection. There was evidence of problems due to waviness in the members of some columns, though this may have been due to over tight tolerances. At least one member was damaged during transportation. The contractor's experience was mainly in small bridges, and though the organization was generally efficient there was a lack of appreciation of the magnitude of the job. The contractual arrangements were unusual for a large structure and the general financial climate surrounding the Quebec Bridge Company was uncertain and 'seriously interfered with carrying out of the undertaking' [5]. There was also evidence of some political pressure when the Dominion government came to be involved, in that it desired that the bridge be opened in time for the Quebec Tercentenary in 1908.

In contrast to the first Quebec Bridge, the second bridge [6] was very carefully researched and designed. It was extremely unfortunate that a failure in one of the castings during the erection of a suspended span caused the loss of that span.

Interpretation of the Assessments

The assessments in Table 1.2 are presented in a form that makes it difficult to draw conclusions about the relative inevitability of each failure and to identify the dominant reasons for the accidents. It must be appreciated that these are subjective assessments and should be interpreted quite differently from the results of objective experiments. There is obviously much more uncertainty in them and they depend crucially on the person making the assessment. They are an expression of engineering judgement. However, in order to reach conclusions from past failures and to produce predictive methods to help avoid future problems, two numerical interpretations of the assessments will be used. In this section a simple scale 0–1 will be used and in Part II (not included in this volume) fuzzy sets were used [37].

Each assessment in Table 1.2 was given numerical values as follows:

1	very high confidence	0.2	A	very low importance	0.2
2	high confidence	0.4	B	low importance	0.4
3	medium confidence	0.6	C	medium importance	0.6
4	low confidence	0.8	D	high importance	0.8
5	very low confidence	1.0	E	very high importance	1.0

For each accident and for each parameter the two assessments were multiplied together to give a value x. The summations of all parameter values of x for each accident are given in Table 1.3 and the summations over all accidents for each parameter are given in Table 1.4. In Table 1.3, summations of x^2 are also given for comparison as this summation tends to

Table 1.3. Results of simple analysis: Failures.

	Simple analysis			
Accident	Σx	Order	Σx^2	Order
Quebec 2	2.52	1	1.45	1
Bedford	2.68	2	1.68	2
Second Narrows	3.12	3	1.90	3
Heron Road	3.68	4	2.04	4
Point Pleasant	3.96	5	2.78	5
Mt Gambier	4.60	6	3.16	6
Camden	4.76	7	3.27	7
Tacoma	5.08	8	3.91	9
Aroyo Seco	5.12	9	3.81	8
Lodden	5.60	10	4.12	10
Ferrybridge	5.92	11	4.34	13
Transocean Ill	5.96	12	4.13	11
Ardeer	5.96	12	4.36	14
Ilford	6.16	14	4.28	12
Listowel	7.00	15	4.99	15
Stepney	7.72	16	5.94	17
Aldershot	7.96	17	6.32	19
Sea Gem	8.12	18	6.30	18
Ronan Point	8.68	19	5.89	16
Kings Bridge	8.68	19	6.77	20
Westgate	11.32	21	9.23	21
Quebec 1	12.04	22	9.29	22
Tay	12.96	23	10.46	23
Example	2.76		0.71	

Table 1.4. Results of simple analysis: Parameters.

Order	No.	Brief description	Size
1	5a	Design error	15.48
2	6c	Construction error	11.88
3	6e	Contractor's staff site control	11.76
4	5d	Designer's staff site control	11.68
5	2b	R&D information	10.88
6	2a	Calculation procedural model	10.68
7	4b	Unknown material effects	9.32
8	4e	New structural behaviour	6.52
9	1a	Overload	6.52
10	5e	Specifications	5.76
11	4d	Step change in structural form	5.36
12	7a	Contractual arrangements	5.32
13	8	Use of structure	5.24
14	4a	Materials well tested	4.52
15	7b	General climate	4.52
16	6b	Sensitivity to erection	4.46
17	5b	Calculational errors	3.8
18	4c	Form of structure well tried	3.24
19	6a	Construction methods well tried	2.8
20	6d	Contractor's experience	2.56
21	5c	Designer's experience	2.52
22	1b	Strength variability	2.48
23	3b	Sensitivity to random hazards	1.92
24	3a	Random hazards	1.28

repress the lower values of x and give predominance to the higher values. The accidents are listed in the order of value of these totals, as these are a crude measure of the degree of likelihood (in these cases inevitability!) of a structure having an accident. The fact that the two Quebec Bridge failures are numbers 1 and 23 in the list bears this out because the number of factors contributing to *the* first *Quebec* failure is far in excess of the unfortunate second Quebec failure.

Table 1.3 also includes a total for an example of a structure which has not failed. This estimate was made by a consulting engineer in practice and serves as a comparison with the totals gained by the failures. It is, in fact, for a four-storey office block of steel-concrete composite construction, and the score is a composite of many small assessments. The total $\sum x^2$ demonstrates this, as the influence of assessments less than 1 is considerably reduced. However, *these* simple totals are poor indicators of the degree of proneness of the structure to failure.

Table 1.4 shows perhaps the most interesting table of results. Again it must be remembered when interpreting this table that the sample of failures taken is not random and includes only failures important enough to merit individual reports of inquiry. Thus, it is not surprising that the scores obtained for parameters 3a and 3b, the random hazards, are very low. A random sample of all failures would produce a very high score for these parameters due to the known high incidence of fires, floods, explosions, etc. Also, similarly it might be expected that parameter 2a would have an increased score if serviceability of partial failures of buildings due to foundation settlement and creep and shrinkage cracking of concrete were included. However, Table 1.4 does have an importance with respect to failures of a major type and shows the predominance of human error in causing these structural failures. Design errors varying from simple second-year undergraduate mistakes, such as the Second Narrows Bridge, to a fundamental misunderstanding of the overall structural behaviour, such as at Ronan Point and Aldershot, are the predominant total. Inadequate site control and checking by the contractor and the clients' representatives are also major parameters. The lack of enough research and development information is an inevitably high total in a list of accident parameters and reinforces the case for an increased expenditure in this field. The uncertainty surrounding the calculations one can perform in making structural design decisions, whether due to loads, materials or the application of theory, is also obviously important (parameters 4b, 4e, and 1a).

These results, of course, may well be affected by the inclusion in the sample of 23 failures of structures of similar type. Failures of falsework, box girders, structures of high alumina concrete or brittle steel, represent over half of the sample. It is interesting to note that two-thirds of the bridges but only one-quarter of the other structures in the sample failed during construction and erection.

Conclusions

A checklist of parameter statements which enable an independent observer of a structural project to make an assessment of the degree of proneness of the structure to an accident can be formulated.

Such a proposed checklist, when used to assess parameters for 23 major structural accidents, shows that failures are due to a variety of causes and combinations of circumstances. However, human error in using existing technology is the predominant overall factor in the accidents considered.

Insufficient research and development information and the resulting uncertainty surrounding design and construction decisions are also major factor in the failures considered.

References

[1] Blockley, D. I. (1974). Structural design decisions and safety. *Pubs. Int. Ass. Bridge Struct. Engng.*, 34, 1–18.
[2] Blockley, D. I. (1977). Analysis of structural failures. *Proc. Inst. Civil Eng., Part I*, 62, 51–74.
[3] Baker, M. J. (1976). *The Reliability of Reinforced Concrete Floor Slabs in Office Buildings: A Probabilistic Study.* Report 57, CIRIA, London, March.
[4] HMSO (1880). *Report of the Court of Enquiry upon the Circumstances Attending the Fall of a Portion of the Tay Bridge on 28th December 1879.* London.
[5] Thomas, J. (1972). *The Tay Bridge Disaster, New Light on 1879 Tragedy.* David & Charles, Exeter.
[6] *Royal Commission on the Collapse of the Quebec Bridge.* Vols. I, II, III (1908). Ottowa.
[7] *Report of the Government Board of Engineers.* Vols. I, II (1918). Dept. Railways and Canals, Canada.
[8] Amman, A. H. *et al.* (1941). *The Failure of the Tacoma Narrows Bridge.* Federal Works Agency, Washington, DC.

[9] *Report of the Royal Commission into the Failure of the Kings Bridge* (1963). Victoria, Australia.

[10] *Collapse of US 35 Highway Bridge, Point Pleasant, West Virginia, December 15* (1967). Report No. NTSB-HAR-71-1, National Transportation Safety Board, Washington DC.

[11] *Report of the Royal Commission into the Failure of West Gate Bridge* (1971). Victoria, Australia.

[12] Freeman, R. and Otier, J. R. H. (1959). The collapse of the Second Narrows Bridge, Vancouver. *Proc. Instn. Civ. Engrs.*, 12, N36–N41.

[13] Health and Safety Executive (1975). *Final Report of Advisory Committee on Falsework.* HMSO, London, June.

[14] Acres, H. G. (1966). *Report on the Heron Road Bridge Failure.* Report to Supervising Coroner of Ontario, Toronto, November.

[15] *Collapse of Falsework for the Viaduct Over R. Lodden on 24th October 1972* (1973). HMSO, London.

[16] *State of California Business and Transportation Agency. Final Report: Investigation into Collapse of Falsework, Aroyo-Seco Bridge Road. 07-LA-210* (1973). Department of Public Works, Division of Highways, January.

[17] NRC Canada (1960). *The Collapse of the Listowel Arena.* Div. of Bldg. Res. Tech. Paper 97, Ottowa, May.

[18] Building Research Station (1963). *The Collapse of a Precast Concrete Building.* HMSO, London.

[19] Clarke, B. L. (1966). *County of Bedford, New County Hall, Report on the Structural Engineering Aspects of the Original Design and Recommended Remedial Measures.* County of Bedford.

[20] Ministry of Housing and Local Government (1968). *Collapse of Flats at Ronan Point, Canning Town.* HMSO, London.

[21] Department of Education and Science (1973). *Report on the Collapse of the Roof of the Assembly Hall of the Camden School for Girls.* HMSO, London.

[22] Bates, C. C. (1974). *Report on the Failure of Roof Beams at Sir John Cass's Foundation and Red Coat Church of England Secondary School, Stepney.* Building Research Establishment, Garston, June.

[23] Mayo, A. P. (1975). *An Investigation of the Collapse of a Swimming Pool Roof Constructed with Plywood Box Beams.* Building Research Establishment, Garston, April.

[24] *Report of the Committee of Inquiry into the Collapse of Cooling Towers at Ferrybridge Monday 1st December 1965.* CEGB, London.

[25] Johns, P. M. and Mottram, K. G. (1968). Investigation into the failure of the Mount Gambier television mast. *J. Instn. Engrs. Aust.*, 40, 117–121.

[26] Ministry of Power (1967). *Report of the Inquiry into the Causes of the Accident to the Drilling Rig Sea Gem.* HMSO, London, October.

[27] *Report of the Committee of Inquiry into the Collapse of the Cooling Tower at Ardeer Nylon Works, Ayrshire, on Thursday 27th September 1973.* ICI Petrochemicals Division, London.

[28] Department of Energy (1975). *Report on the Loss of the Drilling Barge Transocean III.* HMSO, London.

[29] American Society of Civil Engineers (1972). *Structural Failures: Modes, Causes, Responsibilities.* National Meeting on Structural Engineering, Cleveland, Ohio, April.

[30] Feld, J. (1964). *Lessons from Failures of Concrete Structures.* ACI, Detroit, Michigan.

[31] Hammond, R. (1956). *Engineering Structural Failures.* Odhams Press, London.

[32] McKaig, T. H. (1962). *Building Failures.* McGraw-Hill, New York.

[33] Merchant, W. (1967). Three structural failures: Case notes and general comments. *Proc. Instn. Civ. Engrs.*, 36, 499–500.

[34] Pugsley, A. G. (1969). The engineering climatology of structural accidents. *International Conference on Structural Safety and Reliability.* Washington, pp. 335–340.

[35] Pugsley, A. G. (1966). *The Safety of Structures.* Edward Arnold, London.

[36] Scott, G. (1976). *Building Disasters and Failures — A Practical Report.* Lancaster, Construction Press.

[37] Blockley, D. I. (1975). Predicting the likelihood of structural accidents. *Proc. Instn. Civ. Engrs., Part 2*, 59, 659–668.

Chapter 2

Structural Failures and the Growth of Engineering Knowledge*

Abstract

One aspect of the study of engineering failures, which has perhaps been somewhat neglected, is the relationship between them and the growth of engineering knowledge. A modern view of the nature of scientific and mathematical knowledge as being based on models or representations of our experience of the world, and its relationship to engineering knowledge, is presented. Following the ideas of Popper and Kuhn, science and engineering are seen as problem-solving activities within current paradigms. The essential difference between these activities is based on the consequences of incorrect conjectures about the solutions. The primary logical aim of the scientist is to falsify his conjectures as ingeniously and in as well controlled a manner as he is able; whereas the engineer is interested in safe, cautious conjectures which will not be falsified. Engineering failures are the occasions when the conjectures are in fact falsified and they are therefore central to the growth of engineering knowledge. Brief examples of failure which have affected the current engineering paradigm are given.

---

*This chapter was originally published in *Proc. Instn. Civ. Engrs., Part I*, 1980, 68, 719–728 (co-author: J. R. Henderson).

Introduction

It is important that the civil engineering profession, like any other profession, extracts as many lessons as possible from its failures. Enquiries into individual accidents often highlight the immediate technical causes of breakdown and periodic attempts are made at an overall review of the reasons for failure [1–4]. There is another aspect to this problem which in the past has been neglected: the relationship between failure and the growth of engineering knowledge. Philosophers have discussed at length the growth of scientific knowledge, almost certainly assuming that engineering is simply an applied science and that its problems are therefore subsumed under those of science; clearly, this is unsatisfactory. It is the purpose of this chapter to discuss a possible mechanism of the growth of engineering knowledge with failure as an essential component, which is analogous to the explanations of the growth of scientific knowledge as given by Popper [5] and Kuhn [6].

An understanding of this process will also have other benefits. Effective communication between researcher and practicing engineer [7] may be improved if the respective roles of these two groups in the development of engineering knowledge are understood, for instance, many disagreements stem from a lack of consensus view of the role of mathematics and science in engineering. Much disagreement in debates about what should be taught in undergraduate engineering courses flows from a lack of identification of the intellectual status of engineering knowledge. A big research effort may be put into a topic which if seen in a wider context may only have a marginal relevance to the present difficulties in the continuing development of engineering knowledge.

Scientific Knowledge

Science and mathematics are two key areas which form a major part (but not the whole) of engineering knowledge and so it is necessary to begin by briefly discussing them.

Since the turn of the century, there has been a revolution in the way philosophers view science. Before Einstein's relativity theory was accepted, and prior to the development of quantum mechanics, scientists and philosophers thought that scientific knowledge, and in particular Newtonian mechanics, was *absolute* Truth (note the capital T to indicate truth in all contexts). Kant, for example, was strongly under the influence of Newtonian mechanics and thought he had identified all the components of the categories which underpin all knowledge. This view is now untenable. Heisenberg's uncertainty principle, which effectively asserts that there are limits to what we can measure, is an example of the new attitude of uncertainty. In quantum mechanics, a unique correspondence between precise positions and momenta of some postulated element, like an electron, at two different times cannot be established. It is possible only to discuss the problem in terms of probabilities. Modern research in physics has destroyed the hopes of the last century that science could offer us a picture and true image of reality. Science today has a much more restricted objective; it can now only hope to organize experience to enable some sort of prediction. Our scientific hypotheses are man-made devices in which symbols are used to represent features of our immediate experience, as well as defined concepts, which are used by us in a way decided by us. Scientists are free to construct their own systems and to use theoretical terms in them in any way they think most profitable. Thus, science can no longer claim to be the 'truth'; we can treat it *as if* it is the truth but cannot assert that it is true. This leads us naturally to the idea that scientific hypotheses are models; they are models or representations of our experience and thoughts about the world.

In accepting this changed view of scientific knowledge. Popper has rejected the idea that induction is the hallmark of the scientific method. The traditional inductivist view of science is that the process of development of knowledge consists of the following:

(a) observation and experiment;
(b) inductive generalization;
(c) hypothesis;
(d) attempted verification;
(e) proof or disproof;
(f) knowledge.

In Popper's opinion, this view of the power of generalizing from the particular to the general in the scientific method is untenable. He maintains that the growth of scientific knowledge is an evolutionary problem-solving process. The scheme expressed simply is as follows:

(a) problem;
(b) proposed solution — a conjecture;
(c) deduction of a testable proposition;
(d) tests-attempted refutation;
(e) preference established between competing theories.

This scheme is a significantly different view of the growth of science and makes science very much more like other human activities. In science, a problem arises from a breakdown of a previous theory and the proposed solution is a conjecture. This admits the value of any human facility to produce ideas whether by intuition, inspiration or any other creative faculty. The real key to the hallmark of science is whether the conjecture enables the deduction of a proposition which may be tested, either experimentally or by observation of the world. These tests will then either refute the conjecture, in which case it is rejected, or show that the conjecture can be tentatively accepted. The conjecture is not *verified*, it is merely *not falsified.* Clearly, some conjectures, the most powerful ones, will pass many of the tests despite the ingenuity of scientists. Newtonian mechanics was such a conjecture until Einstein suggested relativity theory. Thus, highly tested theories or highly corroborated theories are dependable because they have been shown to be useful in many different circumstances, but they are not necessarily true theories. Popper's view of science is that it is simply an evolutionary process of trial and error, of conjecture and refutation.

If a scientific hypothesis is to be used for prediction, certain assumptions have to be made; in particular, it has to be assumed that the world is regular. This assumption cannot be *logically* justified as Hume pointed out. The problem was neatly summarized by Magee [8]. Just because past futures have resembled past pasts, it does not follow that all future futures will resemble future pasts. The assumption, however, may be psychologically justified; for example, we all believe that the sun will rise tomorrow morning.

The influence of this difficulty concerning the necessary assumption of regularity is highlighted in the problems of the social sciences. Popper [3] summarized this problem thus: '... long term prophecies can be derived from scientific conditional predictions only if they apply to systems which can be described as well-isolated, stationary and recurrent. These systems are rare in nature; and modern society is surely not one of them ... The fact that we can predict eclipses does not, therefore, provide a valid reason for expecting that we can predict revolutions'.

Thus, we can summarize the modern view of scientific knowledge. It is a set of models of the world which are not the Truth but are true in a context. This does not mean that they are not useful; the most dependable hypotheses are those which have been highly tested. Popper has also shown that any measure of this dependability is not a probability [7, 9].

When preference is established between competing hypotheses the chosen ones constitute part of what Kuhn called the current paradigm [6]. Within scientific knowledge there are many hypotheses of different levels of scope and generality, arranged in a hierarchical structure. For example, in structural mechanics Newton's three laws at the highest level are the most general and wide-ranging. Other hypotheses at a lower level vary from the principle of virtual displacements to formulae such as that for the Euler value for the stability of struts. As scientific knowledge develops, clearly the highest and most wide-ranging theories become more and more firmly established as part of the current paradigm. Thus, when a more powerful theory is proposed to replace a high-level hypothesis the consequences are widespread. Einstein instigated a scientific revolution when he proposed relativity as a replacement for Newtonian mechanics. Smaller changes in the current paradigm have a less extensive effect and therefore constitute only minor revolutions. The growth of scientific knowledge depends on the investigation of the current paradigm through controlled testing. There will be periods of 'normal science' when the broad outline of the current paradigm is accepted. As the testing of hypotheses continues and evolves, ideas will change and anomalies will be found but they will be explained with only minor adjustments of the current paradigm. The volume of anomalies may grow and alterations to the paradigm become necessary at higher and higher levels, until only a major revolution will achieve a satisfactory resolution of the problems. This is exactly what has happened in modern physics.

Associated with this process is the human problem of the general acceptance of the new ideas. Inherent in all human systems is an inertia which will manifest itself in a reluctance to accept fundamental alterations to the current paradigm. Clearly there will be people who readily accept such an alteration but the majority will only be convinced after a considerable period of time.

Mathematical Knowledge

A formula or a piece of mathematics which appears in a textbook or a technical paper may represent a piece of science, engineering or mathematics. The principles of deduction may appear to be so similar that it may be impossible at first sight to tell the difference. In fact, the difference lies in our grounds for believing the formula. In mathematics, the axioms stand at the head and everything else is deducible from them in a tight formal deductive system; in effect a theorem of mathematics is a useful restatement of the axioms. By contrast, in a scientific system the high-level hypotheses are justified only to the extent that the propositions deduced from them agree with observations made in the world. Braithwaite used a zip fastener analogy:[10] 'The truth value of truth for mathematical propositions is assigned first at the top and then by working downwards, in a scientific system the truth value of truth (i.e. conformity with experience) is assigned at the bottom first and then by working upwards'.

A modern philosophical view of mathematics is that it is a strict formal language consisting of theorems which express no more than its axioms. It is only when the mathematics is *interpreted* by being used to discuss a scientific model that uncertainty (other than that contained in the validity of the axioms) is introduced.

Just as physics has undergone revolutionary change this century, so has mathematics. In 1930, Gödel wrecked the then existing notions of mathematical proof. He showed that if axiomatic set theory is consistent, there exist theorems which can neither be proved nor disproved, and that there is no constructive procedure which will prove axiomatic set theory to be consistent. In fact, later developments have shown that any axiomatic system, sufficiently extensive to allow the formulation of arithmetic, will suffer the same defect. In fact, it is not the axioms which are at fault but arithmetic itself. Stewart concludes his book [11] '... so the foundations of mathematics remain wobbly despite all efforts to consolidate them .... For the truth is that intuition will always prevail over mere logic .... There is always the feeling that logic can be changed; we would prefer not to change the theorems.'

Thus, even mathematics can be conceived of as a model of the way we think about the world and is therefore open to change just as any other human activity.

Engineering Knowledge

If scientific and mathematical knowledge is perceived as sets of models, then engineering knowledge cannot claim greater status and must also be so characterized. Engineering is also quite clearly a problem-solving process just as Popper describes science, but there are differences which must be explained.

The knowledge of the engineer has been characterized as 'know-how', whereas that of the scientist is said to be 'knowing-that' [12]. This, in the view of many, implies that scientific knowledge is superior. Superficially the difference may seem substantial, but at a deeper level it is not.

Firstly, consider methods of solving problems in the context of the needs and objectives of the problem solver: de Bono [13] relates a particularly appropriate analogy.

Consider a steep valley that has to be crossed. If you are on foot and in a hurry you could run across the flimsy bridge that spans the top of the valley. If you have a car you would use the shorter and stronger bridge that is set lower down the valley wall. If you had a truck you would want to use an even shorter and stronger bridge set nearer to the valley floor. If you want absolute safety and reliability you would descend to the valley floor, cross it and climb up the other side. These bridges of different strengths set at different levels correspond to levels of understanding. You use the bridge or level that is strong enough for your purpose. You do not need to descend to the valley floor every time you want to cross any more than you need to know the molecular structure of albumen in order to boil an egg. If you are in a hurry the long flimsy bridge across the top of the valley might be more practical.

Thus, it is the usefulness of any particular method which is the main interest of the problem solver. It is quite wrong to suggest that detailed explanations of phenomena are better or worse than those which are vaguer. Often detailed explanation adds no more usefulness but does add a false appearance of validity. Vague answers are more likely to include the 'true' answer, but of course the answers need to have enough precision to be useful.

The engineer has quite a different set of problems from those of the scientist. In designing and building a structure, for example, the engineer comes across problems about which decisions must be made. If sufficient data or well-developed models are not available then he must use his judgement and experience to overcome the difficulty. The scientist can organize his problems so that he works on those about which he has conjectures and about which he has deduced propositions which are testable in a fairly precise manner, usually in the well-controlled confines of a laboratory. Thus, in brief, for the engineer, experience and judgement take over when scientific knowledge fails.

Now the way engineers characterize their collective experience (as distinct from each individual's experience which will not concern us here) is by using 'rules of thumb'. These derive from the craft origins of engineering and are really dependable, common sense hypotheses, but of very restricted scope and application. They have developed by a process of trial and error which is very similar to that described in the previous section, but with important differences which are discussed later. 'Rules of thumb' suffer from the major feature of all common sense knowledge which is that while proponents may claim a rule to be correct, they may not be aware of the limits within which it is valid or successful. A rule is most effective when the underlying factors affecting it remain virtually constant, but since these factors are often not identified or recognized then it is incomplete. Scientific knowledge provides us with models which are much more general and therefore have much greater scope and application.

Historically, the first rules to be developed were rules of proportion, based on the geometry developed by the ancient Greeks. Vitruvius and Palladio quote many examples of them [7]. More modern examples of rules are quoted in the engineer's handbook of 1859 [14]. For example, for the deflection of rectangular beams: 'Multiply the square of the length in feet by 0·02 and the product divided by the depth in inches equals the deflection'. For the strength of cast iron girders the rule was: 'the area of the bottom flange multiplied by the depth both in inches, and the product divided by the length in feet, equals the permanent load distributed in tons, allowing the permanent load to be one fourth of the breaking weight'. Empirical formulae such as these fitted to test data obtained in the laboratory contrast sharply with the approach of the French elasticians (scientists and engineers who use the elastic theory of material behaviour) of the period [7].

There are many modern equivalents of such rules. At the most simple level, for example, are the rules for determining the spacing of bolt holes in a steel joint. Other rules, based on some use of mechanics and some laboratory test data, seem authoritative but if the underlying assumptions are examined they bear only a partial relationship to the actual behaviour of the structural element. For example, in order to determine the number and size of bolts required in a moment-carrying steel end-plate connection, a common assumption made is that the joint rotates about the bottom row of bolts and the forces in the rest of the bolts are proportional to their distances from the bottom row. In reality, the problem of understanding the detailed mechanics of the joint behaviour is very difficult because of the many stress discontinuities and unknown load distributions. The method adopted in design works satisfactorily because it produces reasonably safe and economic solutions.

Clearly, many approximating assumptions are found in all design calculations. Joints are assumed to be pinned or fixed, loads are assumed to be uniformly distributed, wind loads are assumed to be static pressures. The approximations are justified only to the extent that they have been used in the past to produce designs which have not failed.

Thus, three components of engineering knowledge have been identified, all of which are models of our collective experience. The first is scientific knowledge, consisting for structural engineers of Newtonian mechanics and tested in the precise confines of the laboratory. The second is the application of these models, with necessary approximating assumptions, to the design of actual structures. These applications are tested by the success or failure of actual structures and by some laboratory work on idealized components. The third component, 'rules of thumb', is also tested by these methods. The nature of this testing is discussed in the next section.

Just as there are current paradigms in science, so there are current paradigms in engineering. In order to design a given type of structure a current set of models are used which make up the current calculation procedure model (cpm). For example, the design of steel pitched roof portal frame buildings involves the use of plastic theory and a number of simplifying assumptions including pinned or fixed joints, ignoring the stiffening of cladding and its effects on the postulated failure mechanism. Among many other assumptions are that the roof loads are taken to be uniformly distributed, and checks are made to account for such effects as finite deflections before collapse, strain hardening and buckling. Clearly, further research will modify this model at many different levels. Some will involve only minor changes; others may involve extensive alterations. Failure of a structure can lead to a major revolution in the cpm.

Failure as a Test of a cpm

The similarity between the growth of scientific knowledge and the growth of engineering knowledge can now be recognized if Popper's problem-solving view is adopted. The scheme for the development of engineering knowledge analogous to that given earlier for scientific knowledge is as follows:

(a) problem;
(b) conjectural solution (by recognition of some approximate similarity with previous problems or by speculation — usually within the current paradigm);
(c) appraisal of the consequences of the solution;
(d) decision on a set of actions (i.e. design);
(e) carrying-through of actions (i.e. construction);
(f) testing of solution (by the performance of the structure: if it fails, it is falsified);
(g) feedback concerning the dependability of the current paradigm.

The components of this scheme are obviously not clear-cut and independent but the essential development is contained within it. The important testing phase is item (f).

The engineer designs and builds in the hope that his structure will operate successfully and will not fail. In using and developing the cpm, engineers are interested primarily in safe, cautious conjectures (while acknowledging economic constraints) because the consequences of failure are so severe. Thus, a cpm is rarely falsified directly in service: instead, great reliance is placed on partial testing of isolated aspects of it using idealized components in the precise conditions of a laboratory. The procedures are only falsified in the laboratory under conditions which do not completely reflect the conditions under which the actual structure operates.

This aspect of the scheme contrasts sharply with that of the scientist because the latter tries to falsify his conjectures as ingeniously as he is able. It is precisely because he can do this that science progresses rapidly. The engineer has no wish to falsify his rules directly and so the boundary of the cpm is only indicated by the occurrence of an accident. The boundary will be very difficult to define exactly because failure will be due to a combination of circumstances: the role of any particular rule in the cpm will not be directly isolable. Design rules are therefore only weakly not falsified. In fact, a particular rule in a cpm may be false but its effect may be masked by the conservative assumptions made in the rest of the design.

The difference between the work of the engineer and the scientist is not therefore a fundamental one, or even a substantial one. It is not due to the different nature of the methods each uses or even the way each perceives the world, but it is *due to the differing consequences of error in the predictions they make*. Engineers do not want to falsify their conjectures; by contrast, this is the scientist's primary aim. Engineers are interested in safe, cautious theories; scientists are interested in detailed, accurate theories. Both are interested in solving problems. Structural engineering scientists tend to be dominated by the scientific interest in accuracy and, as a result, often frown on 'rules' as being intellectually inferior. Designers rely on 'rules' when science lets them down. Many misunderstandings arise because of the failure to appreciate this distinction.

Failures and the Current Paradigm

The discussion so far presents us with a strange antithesis: it is the success of engineering which holds back the growth of engineering knowledge, and its failures which provide the seeds for its future development. Not only is it necessary therefore to identify the particular causes of an accident and the dominant causes of a group of accidents, but it is also important to identify the overall changes in the current paradigm which have followed. This latter aspect is discussed briefly in this section. The changes occur at many different levels with widely varying scope as has been mentioned, and are often difficult to define in any precise sense. Many of the changes mentioned here are well known and have been discussed individually at length. The purpose of this section is briefly to put them in the perspective of the previous discussion on the growth of engineering and scientific knowledge. One of the important consequences of a heightened awareness of the changes is that new developments in research and in practice can be seen as part of the continuing process of the growth of engineering knowledge. This aids an understanding and appreciation of the roles of those who contribute to that process.

The changes are identified briefly in five groups: materials, structural form, theories of response analysis, construction methods and engineering organization. A number of changes which are at present in progress are also identified.

Historically, the development of a new material has led to new structural forms, usually after some time delay, although sometimes new ideas, like pre-stressed concrete, have suffered early setbacks because of deficiencies in material quality. Two separate series of structural failures have led to an increased awareness of the potential deficiencies in materials. Firstly, the failures of the school buildings at Camden and Stepney, and other buildings, led to the banning of HAC concrete. Secondly, failures such as Kings Bridge, Point Pleasant Bridge and Sea Gem showed that brittle fracture and fatigue can cause failure of steel structures if adequate precautions are not taken. These failures have stimulated engineering science research into the mechanisms of the processes, and designers have been made aware of the possibility of other material deficiencies.

The failures of the Tacoma Narrows Bridge and the Ronan Point flats are classics and both showed deficiencies in the current paradigm concerning the behaviour of a structural form. The former accident demonstrated the necessity of wind tunnel tests to determine the aerodynamic stability of slender bridges. Wind tunnel testing now has an established place in civil engineering design, where appropriate. The Ronan Point failure demonstrated the possibility of progressive collapse which, although it is not an unpredictable phenomenon, was not part of the cpm of the time.

The failures of Listowel Arena, of Tay Bridge and at Ferrybridge showed deficiencies in the models of loading for snow and wind. The piling up of snow on one side of the Listowel Arena roof was not considered in the calculations and the figure used for the equivalent uniformly distributed load was too small. The wind loading model for the Ferrybridge cooling towers ignored the fact that there was a group of towers rather than a single one.

The Ardeer cooling tower failure and that of the first Quebec Bridge, as well as some of the box girder incidents, showed up deficiencies in the models used for structural response predictions. The relatively large-scale research programme put into operation after the box girder incidents was one reaction to failure which has generated much valuable information and a new cpm for future designers.

The complex situations surrounding the failures of the Tay, Quebec and Westgate bridges made it almost impossible to isolate single causes of failure. These bridge failures, more perhaps than any others, have demonstrated that human problems of effective organization and communication must be seriously considered as part of the cpm.

This brief identification of some of the paradigm changes after past failures leads naturally to a consideration of the current changes which have not fully filtered through the whole of structural engineering. For example, the realization that many failures are due to human error leads to a greater concentration on checking and control procedures and the inclusion of human error in the cpm. This leads to a reappraisal of risk control through regulations. It has also stimulated further research into types of uncertainty and ways of measuring it. The use of probability theory in structural engineering was almost unheard of 20 years ago, but now with the renewed interest in risk analysis and the prevention of failure it is becoming more and more an established part of the higher levels of some cpms in the current paradigm. Finally, the arguments presented in this chapter demonstrate vividly the importance of the full-scale testing of structures and the potential of proof testing (although the cost of the latter is a serious drawback and has to be recouped).

Conclusions

(1) It has been asserted that all knowledge consists of models of our thoughts and experience of the world. The development of scientific knowledge is an evolutionary problem-solving process.
(2) The development of that part of engineering which is not engineering science is also a problem-solving process which has generated rules which are weakly not falsified.
(3) The essential differences between engineers and scientists is that the former are interested in cautious, safe theories and do not wish to falsify their conjectures, while the latter are interested in falsifying their conjectures as ingeniously as possible. The development of engineering rules is therefore much slower than the development of scientific knowledge.
(4) Failure is an essential component of the growth of knowledge. Clearly this is not desirable in any accidental sense but it points directly to an increased research expenditure on full-scale loading tests.

Acknowledgements

The authors would like to thank the Science Research Council and the Department of Transport for their interest and support of the second author under a Science Research Council CASE studentship.

References

[1] Smith, D. W. (1976). Bridge failures. *Proc. Instn. Civ. Engrs., Part 1*, 60, 367–382.

[2] Sibly, P. G. and Walker, A. C. (1977). Structural accidents and their causes. *Proc. Instn. Civ. Engrs., Part 1*, 62, 191–208.

[3] Blockley, D. I. (1977). Analysis of structural failures. *Proc. Instn. Civ. Engrs., Part 1*, 62, 51–74.

[4] *Structural Failures in Buildings* (1980). I StructE Symposium, April.

[5] Popper, K. R. (1976). *Conjectures and Refutations*. Routledge & Kegan Paul, London.

[6] Kuhn, T. S. (1962). *The Structure of Scientific Revolutions*. University of Chicago Press, Chicago.

[7] Blockley, D. I. (1980). *The Nature of Structural Design and Safety*. Ellis Horwood, Chichester.

[8] Magee, B. (1978). *Popper.* Fontana Modern Masters, London.

[9] Popper, K. R. (1977). *The Logic of Scientific Discovery*. Hutchinson, London.

[10] Braithwaite, R. B. (1953). *Scientific Explanation*. Cambridge University Press, Cambridge.

[11] Stewart, I. (1975). *Concepts of Modern Mathematics*. Penguin Books, Harmondsworth.

[12] Jarvie, I. C. (1972). Technology and the structure of knowledge. Chapter 4 in Mitcham, C. and Mackey, R. (Eds), *Philosophy and Technology*. The Free Press, New York.

[13] De Bono, E. (1978). *Practical Thinking*. Penguin Books, Harmondsworth.

[14] Anon (1859). *Engineer's and Contractor's Pocket Book for 1859*. John Weale, London.

Chapter 3

Design Practice and Snow-Loading Lessons from a Roof Collapse*

Abstract

We describe the collapse of a factory roof in the UK under snow loading. The mode of failure was repeated in a number of similar structures. As discussed in a recent BRE Digest, these failures may have been due to the increased sensitivity of certain types of structure to snow drifting. In this chapter, it is shown that the failure occurred from the unintended consequences of progress in our understanding of the structural behaviour of cold-formed steel purlins. Several lessons are drawn regarding the design of these elements, the role of Code recommendations, the production of safe load tables for design, and the need for engineers to be aware that human action can result in unintended consequences and decisions should be taken so as to minimize their effects.

Introduction

As a result of a number of collapses of roofs under snow loading, the Building Research Establishment (BRE) has recently produced a *Digest* [1]

---

*This chapter was originally published in *Struct. Eng.*, 1986, 64A, No. 3, March (co-authors: N. F. Pidgeon and B. A. Turner).

with new proposals for designing against drifting snow. In this chapter, some lessons are presented which have been drawn from a study of one particular collapse and its background circumstances. The study has been concerned not so much with the technical reasons for failure as with the social and organizational background of the failure. It will be argued that the failure occurred from the unintended consequences of progress in our understanding of structural behaviour. As discussed in the BRE *Digest*, although snow falls in most parts of the UK every winter, occurrence of structural damage due to heavy snow has been rare, until the recent spate of incidents. The Code of Practice for the UK, CP3: "Chapter V.· Part I: Loading: Dead and Imposed Loads" has recently been updated to BS 6399: "Part 1: Design Loading for Buildings", but the design figure for the imposed loading on roofs to which access is restricted is essentially unaltered at 0.75 kN/m^2 uniformly distributed. The BRE *Digest* points to the fact that winds may induce drifting of snow, particularly if there is a step in the roof line or if the roof has a parapet wall. Of course, this possibility of high localized loading has always been present, but the traditional margins of safety apparently have been sufficient to cope. Recent developments in design procedures, however, have led to structures that seem to be sensitive to local loading, and this is illustrated by the following case study.

The Collapse

The collapse was that of an industrial building that had developed in three phases over approximately 10 years. Each phase was designed by the same engineer. Phase I was completed in 1970 and was a single-storey steel structure, 36.5 m (120 ft) span × 45.7 m (150 ft) length and a height to eaves of 4.3 m (14ft 3 in).

The roof was a tapered plate girder supported on three props (Figure 3.1). Lateral loading was taken by bracing in the roof and walls. Cold-formed steel purlins supported asbestos–cement-insulated roof cladding. In the design, an attempt was made to keep foundation forces to a minimum as the site had a history of mining subsidence.

Phase II of the factory was completed in 1975 in response to the client's need to expand his business and was an extension to, and essentially a replica of, the phase I design. The floor area was thus increased by

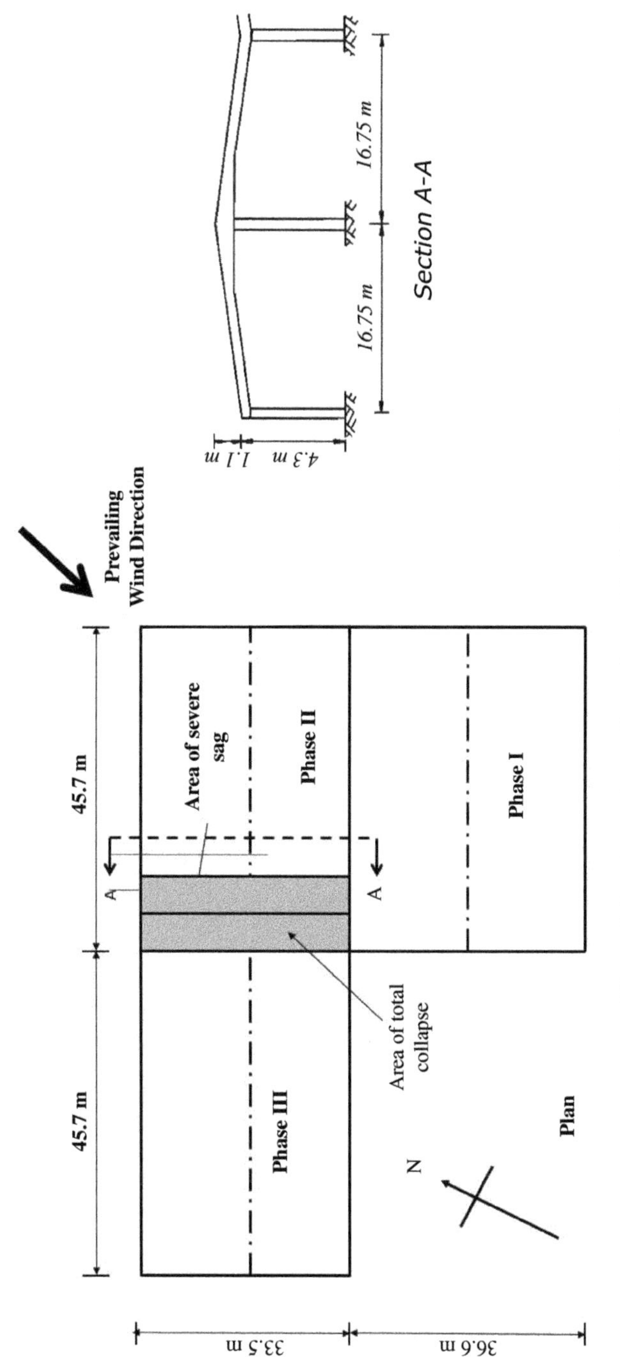

Figure 3.1. Approximate dimensions of the industrial building.

1,530 m^2 (16,500 ft^2). It was quite natural for the client to approach the same practice of consulting structural engineers, but when a further 'end on' extension (phase III) was required by 1980 the practice had dissolved. The client therefore approached the local general contractor who had built phase II. This contractor was small, but well experienced, and with a reputation built up over 25 years of local work. The contractor formed a 'package deal' by engaging an architect and the engineer for phases I and II who now had his own practice.

Phase III was completed in 1980 and again the same design concepts were used to extend by a further 1,530 m^2 (16,500 ft^2). This time, however, the floor area was required for storage, and the roof was approximately 1.5 m higher than phases I and II, because a specific height of storage racking had to be accommodated. There was no reason to suspect any inadequacy in these designs, as phases I and II had been entirely successful. As the engineer who designed all three phases said, "any mathematical error in the designs for phases I or II would have shown up in the differences between them and the calculations for phase III". As mentioned previously, one of the problems uppermost in the engineer's mind was the design of the foundations. Apart from possible mining subsidence, during the construction of phase II the contractor found an extensive layer of peat which continued to be a problem during phase III. The site investigation, which had been provided by the local authority which owned the site, had only mentioned traces of peat. As the engineer commented in hindsight "the loading on the roof was the last factor to worry about really it 'is in the Codes'".

In 1982, during a severe snowstorm, two bays of the roof of the phase II building, adjacent to the phase III building, collapsed (Figures 3.1–3.3). A snow drift had formed against the 1.5 m step between the two buildings, causing high local loading. While the storm was relatively severe, the amount of fallen snow was not abnormal. It was reported by the Meteorological Office that 30 cm of snow fell at a nearby reporting station, with a density of approximately 0.14 g/cm^3. Drifts of up to 1 m were reported on the ground around the buildings. The return period of the snowfall was estimated at about 10 years. The prevailing wind of around 20 kN gusting up to 40 kN was slightly east of north, as shown in Figure 3.1. Thus, the wind had blown some of the snow over the shallow pitch of the roof

Figure 3.2. Totally collapsed bay.

Figure 3.3. Bay with severely sagging purlins.

until it met the obstruction in the roof line and caused the severest build-up of the snow over the area where total collapse occurred. The critical event was therefore the combination of snow and wind which caused drifting and local overloading. It was not possible to predict the return period of this critical combination although it was clearly greater than 10 years. There was no build-up of snow on the roof of phase I, although a cantilever door canopy on the leeward side was heavily laden with snow. Using the Meteorological Office estimates of snow density and other measurements taken on site, the localized loads were of the order of $2\,kN/m^2$.

Under this loading the purlins first failed in bending and then, in the area of total collapse (Figures 3.1 and 3.2), the end connections failed allowing snow into the building. On the other side of the central prop the purlins sagged severely, pulling in the columns, but the joints held and the snow was kept out (Figure 3.3). The central props therefore appear to have mitigated the worst effects of the collapse. Since the structure was all steel, the failure was ductile; it was observed by the factory security guard over a period of several hours, and consequently there were no casualties.

No legal action ensued: the loss adjuster for the insurance company clearly felt that no one was negligent. For all three phases, and in particular for phase III, relationships between the major parties (client, engineer, contractor and architect) were good, and a clear brief had been provided by the client. In general, the standard of workmanship for phase II and the section that collapsed was adequate. However, it was discovered after the collapse that the cladding subcontractor had used hook bolts instead of crook bolts to fix the roof cladding to the purlins. This could have been one extra small factor that might have weakened the integrity of the structure. The engineer had specified the type of cold-formed steel purlin to be used, but added a note that any similar purlin could be substituted. The steelwork subcontractor subsequently asked to use a different type of cold-formed steel purlin. The fixing bolts were the responsibility of the cladding subcontractor and the engineer did not notice, during his visits to site, that the wrong bolts were being used. A minor problem also arose with respect to the cladding subcontractor's ability, early in the contract, to meet deadlines agreed with the general contractor. Effective overall programming requires smooth coordination between the general contractor and the subcontractor, and in this particular case the fortnightly site meetings were insufficient to resolve the problems. Because the subcontractor was not local, high-level informal contacts were by telephone only.

Immediately following news of the collapse, the general contractors visited another building that they had recently constructed which they now realized was similarly at risk. Fortunately, although the roof had noticeably bowed under the weight of snow, it recovered when the snow was cleared. As a result, the owner eventually had his roof strengthened.

Analysis of Failure

The first significant feature of this failure is that, while both phases II and III were satisfactory with respect to snow loading as individual units, a risk had been created because of an interaction between the two which was unforeseen at the design stage. Hence, any analysis of the safety of the two phases separately would not lead to an awareness of a potential for collapse.

As reported earlier, there have been other failures of this type and so this incident can be taken as reasonably typical of structures with cold-formed steel purlins and roof upstands which can cause snow drifts. It seems that many structural engineers at that time would have acted exactly as the engineer for this structure acted. This assertion is also supported by the fact, noted earlier, that no legal action was taken.

It appears therefore that this is a particular example of something more general, which might be termed a 'selective representation' of the safety of a structure. In other words, instead of thinking about safety from every conceivable angle, the problem was considered from a relatively narrow point of view. The effect of the step in the roof was not considered, even though any engineer, if asked, would be able to identify the possibility of a snow drift. There are at least two, not independent, factors which produced this situation. The first is simply the fact that, previously, there had been few failures of this type. Of course, snow drifting on roofs to this extent is relatively rare, and therefore many structures will stand for a number of years before being tested by significant loads. The second factor is perhaps more subtle and is related to the role of a Code of Practice in structural design. The attitude of the designers was apparently that, if a Code specifies certain minimum loading criteria, those are all that need be considered. This is, of course, in one sense at least, quite natural and reasonable because, in a busy practice working under some pressure, the engineer is acting responsibly [2]. The Code represents 'peer group' opinion on the matter, as well as being part of the requirements of the Building Regulations, and so, legally, the engineer is acting with a duty of care. The role of the Code tends to be even stronger, however. The design engineer stated after the collapse that, even with hindsight and the new knowledge gained from such a failure, he would have had difficulty in persuading the client, on purely commercial grounds, that the steel sections at risk would require strengthening beyond the level recommended by the Codes. However, he admitted that he would now at least consider warning the client by letter. Other engineers interviewed about this problem have suggested that, until the Code was formally changed, engineers would inevitably carry on designing structures that are at risk. As one engineer said, "engineers use such Codes for guidance every working day, and what better to use?"

Everybody expressed great surprise at the failure, not least the client. He had initially perceived the structure to be a very substantial one that was 'not built with any expense spared'. Correspondingly, the client blamed an unlikely conjunction of circumstances (wind plus snow). However, in hindsight, the client was adamant that, as the one who had suffered the direct consequences of the failure, he did not want it ever to happen to him again. There was no longer *any risk* that he would be prepared to take.

In order to understand why this situation over snow loading has developed, we need to inquire into the background of technological change that has occurred in the design of the roofs of industrial buildings over the last 20 years. The discussion may then shed light on the continuing role of Codes of Practice in structural design for safety.

Genesis of Hazard Potential

The background to this failure can be expressed in terms of an 'incubation period' [3]. Its starting point was the introduction of a new philosophy, based on load testing, for the design of modern, light-gauge cold-formed steel purlins. Its conclusion was the several roof failures under snow loading. In hindsight, it is clear that this change in methods had put pressure on some of the tacit assumptions incorporated in the then current loading

Code of Practice (CP3: Chapter V: Part 1: 1967), and created what might be termed a new 'hazard category'.

First, let us consider these tacit assumptions. In 1952, CP3: 'Chapter V: Part 1' was revised and the imposed roof loading of 0.96 kN/m^2 (20 lb/ft^2) was reduced to the current figure of 0.75 kN/m^2 (15 1b/ft^2) where no access is permitted other than for cleaning or repair. Although it is not explicitly stated in that Code, it is implied that this nominal load of 0.75 kN/m^2 is at most equivalent to 0.6 m (2 ft) of loose snow. However, in an appendix it is stated that, although the loads are to be regarded as adequate for most conditions, there could well be exceptions where special figures for loading should be used. In the 1967 revision of the Code no mention is made of an equivalent depth of snow although the design figure is unaltered. It seems likely that the figure of 0.75 kN/m^2 (15 lb/ft^2) involves two tacit assumptions:

- firstly, that 0.75 kN/m^2 (15 lb/ft^2) typically represents the loading imposed by a 0.6 m (2 ft) layer of snow on a roof in the United Kingdom
- secondly, either
    - that the maximum snow depth on any significant area of a roof will be less than, or equal to, 0.6 m (2ft) or
    - that the safety margins are sufficient to cope with reasonable variations above this figure, e.g. because of drifting (these margins, of course, derive from the Codes of Practice and latent factors such as the effects of cladding).

Experience, until recently, has been that these assumptions are acceptable. Such forms of assumption are commonly found in Codes of Practice since they are the basis of dependable (and generally conservative) rules of thumb by which difficult and complex problems can be solved efficiently. However, such rules are more than simple design rules; in effect, they are a professionally acceptable mechanism for setting risk criteria. The Codes recommend acceptable risk criteria without *explicitly* stating them. Thus, by adopting a design figure of 0.15 kN/m2 (15 lb/ft^2) for snow loading, the engineer is accepting an *unstated,* but finite, chance of snow overload.

The incubation period for the recent failures began with the use of cold-formed purlins which were designed on the basis of manufacturers' load tables which, in turn, are based on loading tests. For the particular cold-formed steel purlins used in the structure reported here, these tests were carried out in the early 1970s. BS 449 'The Use of Structural Steel in Building' allows the strength of structural steel members to be proved by loading tests. In particular, Addendum No. 1 lays down empirical rules for designing purlins but, of course, these rules make little mention of end fixity or degree of continuity or indeed the yield strength of steel, except as a factor to be used in the proportioning of the lips and flanges. It was quite natural, therefore, for the manufacturers to take advantage of the Code rules and produce a commercial range of purlins with published safe load tables. Full-scale loading tests of each section were carried out and were used as the basis of the design tables. In a typical test, two parallel purlins were supported over their maximum span and were joined by asbestos cement sheeting held down by fixing bolts onto the top flanges. The sheeting was then loaded with a uniformly distributed load both downwards (for snow) and in the reverse direction (for wind uplift). These loads were in accordance with CP3 (Chapter V: Part 1) and the ultimate load factors adopted for the design tables were well within those normally used in ultimate load design. Thus, the design loads for purlins: were within the allowed limits of the existing Codes of Practice for structural steelwork, and resulted in economic savings.

However, this change, from the old empirical conservative permissible stress method for hot rolled sections to a new, laboratory tested, less conservative (but more precise) ultimate load method for cold-formed sections represent a shift in the general philosophy of dealing with acceptable safety margins. These margins, while perceived as acceptable, have nevertheless been reduced on the basis of the more precise response or performance analysis. Of course, safety depends on both the performance and the loading, and neither can be viewed in isolation. Safe reductions in capacity can be achieved only if demand is well specified (or at least not significantly underestimated). In the present case, the change of design philosophy, and consequent reduction in margins, has put pressure on the tacit assumptions underlying the nominal figure for snow loading. In hindsight it is clear that, under certain circumstances (i.e. with some form of roof step or parapet), the 0.6 m (2 ft), uniformly distributed, assumption is not representative of *actual foreseeable conditions*, and this is a situation now critical because of the reduced safety margins. This conclusion depends on a number of reasonable assumptions: firstly, that there have been no underlying changes in weather patterns during the last 30–40 years and that the intensity of snowfalls is unchanged; secondly, that the likelihood of sizeable snow drifts on roofs is also unchanged unless there has been a change in design practice (e.g. more roofs with parapet walls). The recent failures of roofs under snow must therefore be the result of an increased *sensitivity* to overloading. A simple calculation can show how the margins of safety from the ultimate loading tests might be overcome. The tests are performed, quite correctly to the Code, using a uniformly distributed load on the sheeting. This means that each purlin carries one-half of the total load/m of span. Imagine, however, the effect of a triangular distribution of load (a snow drift) across the sheeting between the two purlins with a uniform distribution along the span of the purlins. In this case, one of the purlins would have to carry more load than in the actual test (with a uniformly distributed load over the whole of the sheeting).

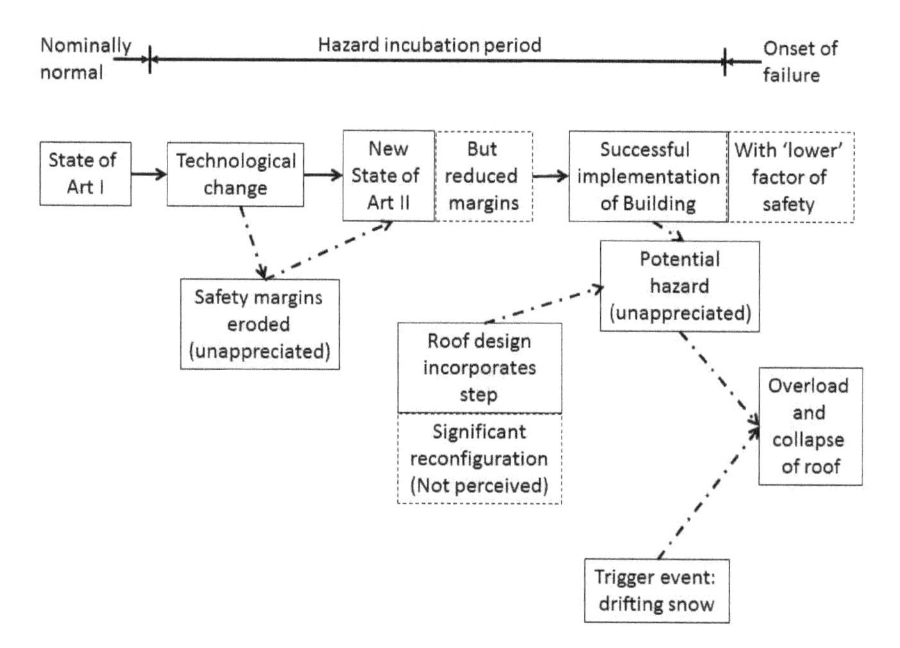

Figure 3.4. Event sequence diagram for roof collapse under snow load.

This would drastically reduce the load factors assumed for the design tables based on such test results. Figure 3.4 is an event sequence diagram summarizing the process leading up to the collapse.

The Lessons

The first lesson from these failures is that, as latent safety margins have been reduced by technically more efficient design, the tacit assumptions in the Code for design snow loading are no longer acceptable. Hence, when designing a roof with a step in it which could result in the formation of snow drifts, a structural engineer should use the recommendations of the BRE *Digest* to determine the design snow loading on cold-formed steel purlins.

The second lesson from this case study is for Code writers. Despite the provision in the Building Regulations that the engineer should adopt higher standards where necessary, minimum acceptable standards written into Codes often become the normal standards because of economic pressures. As noted earlier, the engineer even in hindsight said that, if he had been aware of the importance of the step in the roof at the design stage of phase III, he felt it unlikely that he would have been able to persuade his client to pay for extensive modifications to phase II. Strong commercial pressures are obviously present on the engineer to design as competitively as possible to avoid losing work. Inevitably, therefore, if standard cheaper components are available, which are acceptable under the Codes, they will be used. Furthermore, if a minimum acceptable design loading is stated, it will almost automatically be adopted. This is in effect a risk-benefit decision on the part of the engineer, albeit one that has been deemed acceptable by his peers and is partially sensitive to social control. However, the spirit of a Code (and indeed a requirement of the Building Regulations) is that, although it sets a minimum value, the engineer should also be able to use his engineering judgement to decide in any specific case whether a higher value is necessary. In practice, it is difficult for the engineer to justify to a client the adoption of a standard above the minimum recommended in the Code. An engineer working directly for a client may have more freedom of action in this regard than an engineer working under a design-and-build arrangement. Code writers need to be aware of these problems when setting standards.

The third lesson relates to the production of the safe load tables. The ultimate load tests for cold-formed steel purlins were performed using the Code-specified uniformly distributed loading, and this decision cannot be challenged. However, it is apparent, in hindsight from the case study, that in designing such tests the research engineer has a responsibility to look not only at the Code-specified loading but also at *actual foreseeable loading.* In the present case, this would have entailed dealing with non-uniformly distributed loading patterns. Codes cannot be interpreted entirely in isolation; any particular Code should be regarded as part of a system of Codes which in turn is part of the design decision-making process. An increased understanding of structural behaviour can be utilized in reducing safety margins only if the implications on the loading criteria are understood clearly.

The fourth lesson from the case study relates to unintended consequences, such as the unforeseen interaction between phases II and III. All human action may provoke unintended consequences in addition to the consequences knowingly intended. Although *all* unintended consequences cannot be avoided, some kinds of failure could be averted if the engineer consciously asked 'are there any unintended consequences which might arise from this design decision?' Such consequences, of course, are often difficult to see, however obvious they appear in hindsight, but the possibility of avoiding some kinds of failure makes a regular scan for unintended consequences a worthwhile design habit. The engineering community may benefit from more general liaison with social scientists in connection with such reviews, for philosopher Popper [4], among others, has argued that their function is the study of just such unintended consequences.

These unintended consequences are, of course, part of the more general problem of dealing with uncertainty. The controversy over the introduction of limit state design is a manifestation of the difficulty of dealing with uncertainty in practical design decision-making. In the case study the client was adamant that, in foresight, he would have considered it *inconceivable* that 1.5 m of snow could have accumulated on his roof. Such a belief is congruent with (and possibly tacitly reinforced by) the tendency for the engineer to expect that, if he has designed to a Code, the chance of failure will be zero for all *practical purposes.* Two factors may be relevant to the generation of such an expectation. Firstly, engineers have traditionally not been taught to deal rigorously with uncertainty, either formally through probability theory, fuzzy sets or general uncertainty logics, or perhaps more importantly through the design philosophies recommended in Codes, which have fudged the issues [4]. Secondly, there is a natural human response of defensive avoidance [5]. As Fischhoff *et al.* have noted [6], 'professionals often assume enormous responsibility for other people's lives and safety. They may have to daily give others such assurance as ... that structural member will hold until the other ones are in place. Bearing this responsibility may require a special ability to deny or tolerate uncertainty. In reassuring others about the quality of their decisions they may also be reassuring themselves'.

The final lesson of this case study, and indeed perhaps of all engineering failures, is that those who educate and train engineers should cover the problems of decision-making under uncertainty rather more rigorously than in the past. The state-of-the-art of this topic is in its infancy and has sometimes, in textbooks, been identified solely with subjective probability and decision theory which, of course, is much too narrow an interpretation of the topic. While quantitative techniques are being developed, there is a need for a much more rigorous treatment of design philosophies [4], their relationship with professional responsibility [2], the treatment of unintended consequences, and, at the most general level, with the role of 'risky' technology in society [7].

Conclusions

The failure of cold-formed steel purlins in a factory roof because of snow overload has been described. It seems that this failure was typical of a number of such incidents. The lessons drawn can be summarized briefly as follows:

(1) Structural engineers designing cold-formed steel purlins using safe loads based on ultimate load tests should take account of the recommendations of BRE *Digest 290.*
(2) Code writers should remember that minimum acceptable standards written into Codes almost certainly become the normal standards through economic pressures.

(3) Safe load tables ought to be produced from analysis of, or tests on, actual foreseeable loading (as far as is possible), rather than only on Code-specified loading.
(4) When taking any decision the engineer should try to identify, and minimize, the risk of the unintended consequences of the decision.
(5) Those responsible for the education and training of engineers should cover the problems of decision-making under uncertainty much more rigorously than has traditionally been the case.

Acknowledgements

The work was carried out with financial support of the Joint Committee of the Economic & Social Science Research Council and Science & Engineering Research Council. We are grateful to those who must remain anonymous who freely gave of their time, knowledge, and experience, to help us. We thank C. J. Judge of the BRE for helpful comments on snow loading.

References

[1] BRE (1984). *Digest 290: Loads on Roofs from Snow Drifting against Vertical Obstructions and in Valleys.* Garston, Watford, Building Research Station, October.
[2] Blockley, D. I. (1985). Reliability or responsibility? *Struct. Saf.*, 2, No. 4, 273–280.
[3] Turner, B. A. (1978). *Man-Made Disasters.* Wykeham, London.
[4] Blockley, D. I. (1980). *The Nature of Structural Design and Safety.* Ellis Horwood, Chichester.
[5] Janis, I. L. and Mann, L. (1977). *Decision Making.* Free Press, New York.
[6] Fischhoff, B., Lichtenstein, S., Slovic, P., Derby, S. L. and Keeney, R. (1981). *Acceptable Risk.* Cambridge, Cambridge University Press.
[7] Collingridge, D. (1980). *The Social Control of Technology.* Open University Press, Milton Keynes.

Part II

Joining-Up Theory and Practice

Preamble

One particular case history we looked at in Chapter 1, which was of particular interest and personal concern to me, was the collapse of the Westgate Bridge in Melbourne Australia in 1973. The reason was simply that an undergraduate contemporary of mine was working on site as a Section Engineer for Freeman Fox and Partners — the structural designers. As I recall *Construction News*, a specialist publication, carried transcripts of the official inquiry. I read verbatim the difficult questions and answers put to those giving evidence. For example, how would I have answered 'How flat is a flat plate? I felt keenly that I could have been him. Instinctively I knew 'there for the grace of God go I'. How would I have coped? I saw vividly the effect of unintended consequences of collective and professional decision-making leading to loss of life.

In the media, people talk a lot about rights — perhaps the ultimate oft quoted and controversial example is the European Convention of Human Rights (June 2019).[a] There tends to be much less talk about the flip side, i.e. responsibilities. Like many of the failure case histories we looked at in Part I, the story of the Westgate Bridge[b] exposed the responsibilities of professionals. At that time it appeared to me to be an open wound — there was little health and safety legislation and individual workers found it very hard to do much to influence a situation they thought might be unsafe.

Of course, what constitutes a human right is arguable but will include freedom, justice and peace. Certainly it should include personal safety. As I was thinking about these issues I came across an article by Tom Settle [1] which set me thinking about the notion of responsibility and the legal duty of care of all professional under the law of tort.

The notion of the inductive reliability of a theory or hypothesis, argued Settle, should be replaced by the notion of a responsibility to act on the theory or hypothesis. The taking of responsibility implies not that one has earned the right to be right, or even nearly right, but that one has taken what precautions one can reasonably be expected to take against

[a] See http://www.echr.coe.int/Documents/Convention_ENG.pdf (Accessed on June 2019).
[b] See http://www.westgatebridge.org/ (Accessed on June 2019).

being wrong. The responsible engineer or scientist is not expected to be right every time but he is definitely expected never to make childish or lay mistakes. The latter is negligence. However, *being wrong is not the same as being negligent.*

I saw that it is possible to draw an analogy between the research scientist's search for truth (here I mean a common-sense interpretation of truth, simply as 'correspondence to the facts') and the engineer and applied scientist's search for reliability. Both objectives are qualities of what is being created; for the scientist, knowledge; for the engineer, an artefact. The pure scientist tries, as far as he is able, to control the environment in which he conducts his search (laboratory conditions) and if he follows Popperian logic he sets up bold conjectures and tries to refute them. The applied or engineering scientist likewise attempts to work in laboratory conditions but is often faced with having to produce theoretical models for use by engineers in design. An attempt is made to make sense out of data concerning incompletely understood phenomena in order to help make some sort of prediction.

The deterministic treatment of engineering calculations has its roots in the ideal of 'exact' science. Modern physics has destroyed the idea of science as the absolute Truth (with a capital T to denote truth in all possible contexts). However, the notion of there being an objective truth is tenaciously held by most people — especially engineers. To most of us, it is a matter of common-sense reality — we deal with facts. No one would seriously dispute that $2 + 2 = 4$ or that if a man let go of a stick in his hand then it will fall to the floor. But if one looks into the philosophy of science behind such strict truth we find that it is elusive. In brief, we have to describe a context for any statement we make and then that context needs a context, and so on. In philosophical terms we need a meta-language or a meta-theory.

So, if the engineer cannot rely even on science to provide the 'truth' the problem of ensuring an adequate product seems to become overwhelmingly difficult. Not only is engineering judgement required for the matching between engineering theories and the actual product but also for the assessment of the dependability of the theory itself. I realized that Settle's suggestion is exactly what happens in practice, but no one seems to recognize that. The key idea at the root of all professional and public decision-making is the legal 'duty of care' under the Law of Tort.

So, in Chapter 4, my next **Learning Point No. 4** is that 'we should replace the reliability of an idea with a responsibility to act on that idea'. The ultimate test of any theory or method or technique is that to act on it is a responsible thing to do — not whether it is true or false. The key point is that the responsible engineer takes all reasonable precautions against being wrong and is careful, informed and above all self-critical. Judgements of reasonableness are made by the peer group whose judgements in turn ultimately depend on their ethics.

The argument presented directs attention to the idea that measures of risk are not absolute measures but aids in the process of the management of knowledge to control risk.

I first met Priyan Dias in 1989 and subsequently he spent 10 months in Bristol on study leave funded by a Commonwealth Fellowship in 1992/1993. We have corresponded and collaborated ever since. When Barry Turner suggested I might be interested in the idea of reflective practice introduced in a book by Donald Schön [2] — I was immediately taken with the idea. When Priyan came to Bristol, he and I began to talk it over and together we wrote the paper which is Chapter 5. Later, I came across the work of the American pragmatist philosopher John Dewey who wrote extensively about education. In 1929, he also wrote about certainty and uncertainty[c] but he was rather out of step with the philosophy of his time. Thought and practical action are intimately related he said. He expressed what we tried to communicate in our paper when he said, 'We do not learn from experience … we learn from reflecting on experience'. Priyan and I maintained that systems thinking is the key to reflective practice. We argued that our academic institutions are dominated by the scientific method which, in essence, is analytical and reductionist. We wanted to stress synthesis and a holistic approach without rejecting reductionism.

So, my next **Learning Point No. 5** is that 'academic institutions are currently dominated by technical rationality' — and technical rationality is dominated by the scientific method. One of its important characteristics is a process of selective inattention to make the problems tractable. In Chapter 5, we argue that reflective interaction is a much better framework for understanding and developing a theoretical base for professional

[c] See https://archive.org/details/questforcertaint032529mbp.

practice. However, in 2019 the scientific method still dominates the thinking of academic engineers.

One of the greatest hazards of being an academic engineer (irrelevant mechanic!) is balancing these esoteric flights of fancy with the need to keep 'my feet firmly planted on the ground'. One of the ways I have tried to do this all through my working life was by getting actively involved in The Institution of Structural Engineers (IStructE). I chose the IStructE for two principle reasons. Firstly, the test for professional competence is a tough one — a 7 h written examination in which the candidate has to design a structure according to a given brief. Passing it gives you some practical credibility and kudos. Secondly, by being involved I would meet and talk on a regular basis about the issues of the day that concern practitioners. I was lucky on both counts. Despite my limited practical design experience, I managed to pass the IStructE professional examination to become a Chartered Structural Engineer in 1973. My friends and colleagues who are and have been members of IStructE in the Western Counties Branch have been a pragmatic rock on which I have relied substantially. In particular, Stewart Craddy appointed me as a consultant to his business and I learned a great deal from contact with him and his partners — especially Simon Pitchers. Both have become good friends. The case history of Chapter 3 came from one of Stewart's jobs.

Being President of the IStructE was one of the best years of my life. Chapter 6 is my Presidential address. The title, as I write in the chapter, came to me when I was travelling by train and I overheard the following snippet of a conversation between two fellow passengers: 'We wanted someone who could think out of the box so we didn't want an engineer'. I bristled inside when I heard that. However, on reflection I had to accept that the cliché may be true for some of us all of the time and for many of us some of the time.

Of course, the idea of someone being able to think outside the box is the idea of innovation — creating something new or breaking new ground. In my experience, engineers do this all the time but there are two problems. Firstly, they do not see, and therefore rarely claim, that what they are doing is innovative. Secondly, and perhaps consequentially, others do not recognize engineering innovation either. I began a process that later led to me realizing that engineering has an image problem. So, my

Learning Point No. 6 is that 'innovation is key to future success — we need to think outside of the box'. What is more engineers need to show how they think out of the box when they do. Examples of engineering innovation abound — just think of transport — the railway, aeroplane and car, just think of computers — the mobile cell phone and the personal computer. All of these things are engineering innovations. Certainly, as we face the challenges of the 21st century we will need to be innovative.

References

[1] Settle, T. W. (1969). Scientists: Priests of pseudo-certainty or prophets of enquiry? *Sci. Forum*, 2, No. 3, 21–24.

[2] Schon, D. A. (1983). *The Reflective Practitioner: How Professionals Think in Action.* Temple Smith, London.

Chapter 4

Reliability or Responsibility?*

Abstract

Engineering failures are often due to human fallibility. Are some of the reasons for this very fundamental? Is engineering education to blame or are there fundamental limitations on our ability to use experience, necessarily gained with the benefit of hindsight, to control complex technology? What should society expect of its engineers as regards the safety of structures?

Clearly, these questions are problematical, and many alternative theses may be advanced. Safety analysts would generally agree that society should not expect absolute guarantees of safety. The conjectures advanced in the chapter are (i) the notion of a responsibility to act upon a hypothesis is a fundamental concept; (ii) the measure of the quality of an engineering hypothesis should be 'dependability' not truth; (iii) all measures of risk are not absolute measures but aids in the process of managing knowledge to control risk and, more generally, technology itself. Responsibility and reasonable decision-making are concepts used in the law courts. There is a possibility that total reliance on peer-group approval for the definition of these concepts could lead to stagnation. A deeper understanding of the processes involved could well come from a study of changing paradigms (analogous to Kuhn's paradigms for science) in engineering. Safety analysts should perhaps take a lead

*This chapter was originally published in *Struct. Saf.*, 1985, 2, No. 4, 273–280.

in setting up a dialogue between themselves, engineers and lawyers, in order that the courts appreciate the fundamental/imitations of safety prediction and control.

Introduction

Are engineering failures really failures of engineers? Are the underlying reasons for failures just 'Acts of God', technical errors of some kind, or human mistakes?

Clearly individual cases merit individual enquiry to establish what went wrong and the lessons to be learned. However, a general conclusion which is emerging from research studies on a number of failures is that human error is a major factor. Of course, in a sense, in any engineering project all error is human error because it is people who plan, design, produce and use the product. Engineers and technologists in their quest to discover ways of organizing nature and flushed with their successes in the physical sciences, have perhaps rather neglected the extent to which they rely on human infallibility.

So, given that failures are something to be avoided if at all possible, what needs to be done? Some would argue that our educational system is at fault. A recent CNAA (UK Council for National Academic Awards — abolished in 1992) report, not concerned with failures, on Goals of Engineering Education in the UK, suggests that the education engineers receive is 'both technically narrow and narrowly technical'. Engineering education over-emphasizes engineering science in one field rather than engineering as a process and there are few attempts to teach business and management skills. Thus, if a contributory cause of a failure is, for example, poor project management or a lack of technical 'overview', then perhaps by altering our university and polytechnic degree courses, failures of this type will become less likely.

What Can Society Expect?

However, what is taught reflects, on the whole, our collective understanding of 'things as they are'. It is often argued that it is not the topic of an individual's education which matters but rather his abilities and attitudes. It is not only knowledge and technical ability but also such characteristics as openness of mind, ability to communicate, to organize and to formulate problems which are related to the topic studied in depth. Perhaps, it is not so much what is taught but how it is taught which is important in engineering education.

As far as failures are concerned, perhaps a rather more fundamental question that engineers and scientists should ask of themselves is 'what is it that society can expect of us?' That is, what does society have a right to expect and what are engineers obliged to provide? Should society expect no failures? Is it reasonable to expect perfect or near perfect reliability of an engineering product whether washing machine, bridge, nuclear reactor, offshore oil rig or, most controversially of all, nuclear missile defence system?

It is possible to draw an analogy between the research scientist's search for truth (here we will adopt a common-sense interpretation of truth, simply as 'correspondence to the facts') and the engineer and applied scientist's search for reliability. Both objectives are qualities of what is being created; for the scientist, knowledge; for the engineer, an artefact. The pure scientist tries, as far as he is able, to control the environment in which he conducts his search (laboratory conditions) and if he follows Popperian logic he sets up bold conjectures and tries to refute them. The applied or engineering scientist likewise attempts to work in laboratory conditions but is often faced with having to produce theoretical models for use by engineers in design. An attempt is made to make sense out of data concerning incompletely understood phenomena in order to help make some sort of prediction. The strategy adopted may range from that of the pure scientist at one extreme to mere curve fitting to data points at the other. The engineer uses many different theoretical models with varying sets of idealizing assumptions concerning the quality of the matching between the theory as developed in laboratory conditions and the reality of the World Outside the Laboratory (WOL). At the most basic level, Newtonian mechanics is assumed to be an infallible description of the physical world. Of course, it is known that there are situations in which this theory is inadequate, is false, but the engineer knows equally well that those situations will not occur in his project. However, for less basic theories with sets of complex assumptions (for example, the prediction of fatigue damage in bridges) it is not as easy to be certain. There are at least two difficulties. Firstly, the phenomenon itself may not be well understood and secondly the matching of laboratory tested assumptions with the practical reality may be poor. The importance of these two factors varies across engineering industry from mass production to 'one-off' construction [1].

Construction and Manufacturing

The distinction between 'one-off' and mass production may sound rather trite but it leads to profound differences in the quantity and quality of information available to the engineer. If a product is to be mass produced it makes economic sense to test one or more prototypes; in fact, prototype testing becomes an essential phase of the design and development of the product. By contrast it is uneconomic to test a 'one-off' product to destruction and to use the information gained to rebuild. Thus, the designer of a 'one-off' product obtains much less feedback about the performance of the product in the WOL than does his manufacturing counterpart. The resulting uncertainty largely surrounds the quality of any model, whether scientific or not, that the engineer uses to make his decisions.

This modelling or system uncertainty is due to the lack of dependability of a theoretical model when used to describe the behaviour of a proposed product assuming a precisely defined set of parameters describing the model. This is complemented by parameter uncertainty which is due to the lack of dependability of theoretical propositions concerning the parameters of a theoretical model used to represent a proposed product, assuming that model is precise. In engineering problems where prototype testing is thoroughly carried through, the system uncertainty is much reduced and the parameter uncertainty is dominant. In 'one-off' engineering, both types of uncertainty are very important and in some cases (e.g. geotechnics) the system uncertainty is dominant.

Reliability theory as developed for industries where system uncertainty is small (for example electronic control) must only be applied with care to engineering problems where this type of uncertainty is large. Indeed, it can be argued that probability theory should not be used as a measure of system uncertainty [1].

Decision-Making Using Dependable Information

'The scientific process has two motives: one is to understand the natural world, the other is to control it. Either of these two motives may be dominant in any individual scientist; fields of science may draw their original impulses from one or the other'. C. P. Snow was discussing the relationship of scientist and engineer. He continued, 'The more I see of technologists at work, the more untenable the distinction has come to look. If you actually see someone design an aircraft, you find him going through the same experience aesthetic, intellectual, moral — as though he were setting up an experiment in particle physics'.

Indeed, if one uses Popper's explanation of the growth of scientific knowledge, the similarity between scientists and engineers as problem solvers is very great. The difference really concerns the motivation for solving the problem and the feedback from testing the solution. It has been argued that in a Popperian sense engineering failures are central to the growth of engineering knowledge [2]. However, because the causes of engineering failures are difficult to establish precisely, the relationship between engineering conjectures and their falsification through failure in the WOL, is ill defined and fuzzy. Thus, engineering conjectures are only weakly falsifiable. Indeed, whilst a scientist may wish to conjecture a solution to his problem and then falsify it as ingeniously as he is able in his pursuit for truth, the engineer has no wish to do that at all: falsification means failure, disaster and loss of resources. It is not surprising, therefore, that engineering knowledge of the WOL advances much less quickly than the scientific knowledge of well-defined and controlled situations.

The central activity of both scientist and engineer is that of a decision maker. Whatever hypotheses are conjectured, whatever the problem faced and whatever the motivation of the problem solver, decisions must be taken on the basis of dependable information in the pursuit of knowledge or quality of product.

The deterministic treatment of engineering calculations has its roots in the ideal of 'exact' science. This conception of science as leading to the 'truth' has been destroyed. The prevailing scepticism amongst philosophers about the possible achievements of science has to be contrasted with the apparent success of technology. A number of recent philosophers have tried to resolve this problem of science. For example, Lakatos argues that the scientist aims at a cumulatively fruitful research programme. Kuhn is perhaps more concerned with the sociology of the changes within science which lead to new theories being accepted. Feyerabend argues the rival theories are incommensurable and no common sets of values for arbitration between them exist. Carnap tried to replace the lost certainty of the last century by a measure of justification based on mathematical probability. He argued that it is meaningful to talk of the probability of the truth of a proposition over a finite and determinate interval. The use of this idea together with the interpretation of mathematical probability as a degree of belief (as argued by de Finetti, for example) has been the basis of decision theory and the initial treatment of system uncertainty in reliability theory.

The clash between those following Carnap and those following Popper has been amply discussed by Lakatos [3]. He showed that there is a basic confusion between the notion of the use of probability theory as a 'rational betting quotient' and as a 'degree of evidential support'. Cohen [4] has also destroyed the idea of using mathematical probability as a measure of the dependability of inductive evidence and has suggested a new inductive probability measure, quite different from mathematical probability, for application to legal questions.

It is suggested that what really matters to an engineer is the dependability of a proposition. If a proposition is true it is dependable but if a proposition is dependable it is not necessarily true. Truth is sufficient condition but not necessary condition for dependability. Einstein demonstrated that Newtonian mechanics is not 'true' but is dependable under certain conditions. Repeatable testing of propositions deduced from Newton's laws have shown that they correspond (within defined error bounds) to the facts (are true) but not always. In other words, the truth content of Newtonian mechanics is high even though in a strict sense the laws are false. They are highly tested, highly corroborated and therefore inductively very reliable but the logical probability of their truth is zero. In solving an engineering problem, a whole hierarchy of theories may be used and each theory is only applicable under certain conditions (e.g. elastic behaviour of a material). Even under those conditions the dependability of the use of the theory may not be high (e.g. elastic behaviour of a sub-soil) and it is this that constitutes part of the system uncertainty. The sufficient conditions for dependable information have been discussed in detail [1]. A conjecture is dependable if (i) a highly repeatable experiment can be set up to test it, (ii) the resulting state is clearly definable and repeatable and (iii) the value of the resulting state is measurable and repeatable and (iv) the test is successful. These are sufficient, but not necessary, conditions because the proposition may not be false even though it is not possible to set up repeatable experiments. Deficiencies in any of the ways in which the propositions can be tested or inductively applied obviously leads to uncertainty and a consequent loss of dependability.

Responsibility

If the engineer cannot rely even on science to provide the 'truth' the problem of ensuring an adequate product seems to become overwhelmingly difficult. Not only is engineering judgement required for the matching between engineering theories and the actual product but also for the assessment of the dependability of the theory itself. Settle [5] has tried to resolve this problem by suggesting another criterion, which has great appeal because it is in effect adopted by engineering practice. His suggestion is made in the light of the critical method in science, which is the development of Popper's philosophy of trial and error, of conjecture and refutation, the concept of critical discussion leading to progress. The notion of *the inductive reliability of a theory or hypothesis*, argues Settle, *should be replaced by the notion of a responsibility to act on the theory or hypothesis*. The taking of responsibility implies not that one has earned the right to be right, or even nearly right, but that one has taken what precautions one can reasonably be expected to take against being wrong. The responsible engineer or scientist is not expected to be right every time but he is definitely expected never to make childish or lay mistakes.

Engineering involves decision-making on the basis of information of varying inductive applicability or dependability. A decision may be viewed as a suspension of criticism for a moment. Good decision-making is not a static process, however, and criticism of the consequences must therefore be continued after the moment of the decision. Responsibility, it is argued, is a more useful concept than reliability. In one sense this proposition is acceptable because it points directly to the role of the individual and his duty to work with care and diligence. It is a concept used in law where under the law of tort, for example, the standard of care is that of the reasonable practitioner or one who is careful, informed and self-critical. In another sense, however, the concept of responsibility is difficult to accept. It is difficult to define precisely and it varies very much with individual circumstances. A person holding himself out as possessing a particular skill (a university researcher undertaking specialist consulting work) must exhibit the degree of knowledge and skill reasonably to be expected of any other similar person, but he must neither be expected always to exhibit the very highest degree of skill nor to anticipate and avoid every possible future risk inherent in the particular task he is carrying out.

The definition of what is a reasonable action must in the end be made by the peer group of the person concerned. For the average engineer, one peer group is the membership of the appropriate professional institution. For the detail of a design procedure, a particular code of practice may be the standard of reasonable design practice and must therefore be interpreted as a measure of peer group opinion.

Judgements of reasonableness by the peer group must depend ultimately on their set of values or ethics, concepts which are even less precise. It could be said therefore that the argument has degenerated from the scientific precision of the 19th century to a social scientific vagueness about ethics. That may be so, but nevertheless what has been made clear is the fundamental responsibility of the individual to be diligent and careful, and to take part in the development of the values, the ethics, and hence the opinions of the peer groups of which he is part. The argument helps to clarify one aspect of the role of the professional institutions, in a way that many who are active in their work may feel intuitively to be correct. It resolves the tension between so-called scientific precision and engineering practice. It provides a framework of ideas in which the research engineering scientist and the practical engineer can better relate to each other. It points the way to more research on methods by which the quality or dependability of information and inductive evidence can be judged more effectively, such as Cohen's inductive probability [4] or Baldwin's fuzzy logic [1]. It points the way to the concept of a management of knowledge to control risk rather than a calculation of some absolute measure. There remain, however, two important difficulties which must be resolved. The first relates to innovation and the second to the law.

Responsibility and Stagnation

Consider a particularly creative engineer or scientist, who has a new, rather adventurous, idea. He collects enough evidence to convince himself of the validity of his idea but he is unable (if indeed he tries) to convince his peer group. Nevertheless, he goes ahead with the scheme. The question now is whether or not he is acting responsibly? The simple answer, by the argument presented, is that he is not. For example, Thomas [6] when discussing common causes of liability and their avoidance, states, 'It is appreciated that the following will be unpopular advice, but the fact remains that the best way to avoid litigation is to design conservatively, using trusted materials and methods'. Of course, if the project is entirely successful there will be no difficulty, new evidence will be created. However, if there is an engineering failure, if his idea is refuted, then he could well be sued for acting irresponsibly. This possibility is quite naturally a strong disincentive to the creative engineer. The argument presented so far could thus lend to stagnation and prevent progress. This contrasts strongly with the Popperian view that progress results from bold conjectures ingeniously refuted. A resolution of this problem, which actually faces all 'expert' decision makers, must depend on the possible consequences of a decision. It is clearly irresponsible to make a large step change from current practice, if the possible consequences could be serious in terms of environmental damage, loss of life or resources. However, progress must depend on new development. Responsible decision-making is perhaps therefore an argument for evolutionary change, rather than revolutionary change.

Kuhn has argued that most scientists spend their time doing 'normal science'. Much of the success of this enterprise derives from the scientific community's willingness to defend the current assumptions about the basic nature of the world. Normal science, for example, often suppresses fundamental novelties because they subvert basic commitments. Inevitably, however, anomalies occur which cannot be explained, and the pressures created by these build up until inevitably there is a shift in commitment — a scientific revolution. Kuhn's paradigms of normal science have two fundamental characteristics, firstly, this basic commitment to a set of fundamental ideas and secondly, a set of open problems which the group of practitioners must resolve.

Just as scientists have their paradigms of normal science so, it is possible to argue, do engineers. The nature of these paradigms will, however, depend not only on scientific knowledge but also on other factors such as materials and production techniques. A discussion of these engineering paradigms is beyond the scope of this chapter and it will be sufficient here to adopt an intuitive analogy with Kuhn's paradigms of science. Using this analogy, it is clear that engineers can be creative and inventive and obtain peer-group approval, if they remain within the current paradigms. They must, of course, exhibit the necessary degree of skill if they are not to be accused of acting irresponsibly should something subsequently go wrong. What then of engineers who wish to step outside the bounds of the current paradigms? If the possible consequences of their ideas are not serious there is no problem. If the consequences may be serious there is a dilemma. On the one hand if the development of the idea is prevented progress will be inhibited; on the other hand, sufficient resources must be provided in order to collect sufficient evidence to convince the peer group that the risk is acceptable. This second course of action inevitably takes time and money but if ignored could lead to a failure.

As the power of technology increases so our responsibilities increase. At a more general level of responsibility, Collingridge [7] has argued that the search for better ways of forecasting the social impact of technologies is wasted and that some way must be found to retain the ability to exercise control over a technology. He discusses the notorious resistance to control which technologies achieve as they become mature and the factors which decision makers should consider.

Responsibility and the Law

It was earlier stated that the concept of responsibility is central to the law. Clearly legal practice will differ in various countries but certainly in the UK the legal position is unclear. It will be argued in this section that engineers must more forcibly state their problems and difficulties to lawyers.

Numerous decisions in the English courts have led to persons, who suffer as a result of defects in construction work, suing not under contract law but under the law of tort. If the engineer is not held to have warranted the quality of this work, he will be subject to the duty to use reasonable care and skill; i.e. a duty not to be negligent as discussed earlier. However, there are two major difficulties that the courts must face in resolving disputes. The first is that complex technical problems may be at issue and experts in disagreement; the second is that the courts must distil these complex problems down to a point of law.

An example of the first difficulty is the law of limitation in the UK. Construction defects very often do not become apparent for many years, when original details may have been forgotten, records destroyed or companies out of business. Experts may disagree about the nature and cause of the defects and when they should reasonably have been discovered. The law is clear about the limitation period, six years from the date of the accrual of the cause of action; but the interpretation of the law in deciding when the six-year period commences is problematical. Hence, at a time when the construction industry is in recession, the legal profession dealing with construction is a growth area.

The second difficulty that the courts face in resolving disputes is, in a sense, even more problematical. Judges are not trained to follow non-legal technical argument unless it is reduced to simple terms. There must be a temptation for the judge, in hindsight, to feel that if he can understand the essence of the case then it should have been foreseen by defendants whose business it is to do so. The distinctions drawn in judgement are often very simple. For example, in a recent Court of Appeal Lord Scarman relied on a passage in a 1940s judgement of Du Parcq which distinguished between situations where only services were supplied and those where a chattel was ultimately to be delivered. Lord Scarman concluded that 'one who contracts to design an article for a purpose made known to him undertakes that the design is reasonably fit for its purpose'. Thus, it is possible that the courts may impose a strict liability of fitness for purpose. Here the designer is liable for failure of a structure to achieve its required purpose even if he has used all reasonable care. In other words, he must guarantee absolute safety in law.

All safety analysts and most engineers will understand the absurdity of this requirement. It is beyond the scope of this chapter to discuss the legal difficulties in depth, but it does seem that there is a problem in relying on the legal process to define reasonable and responsible behaviour. It cannot be in the interests of the construction industry to continue a system which, when defects appear, results in long squabbles which benefit no one but lawyers and expert witnesses. It is at least arguable that engineers concerned with safety and reliability analysis should open a dialogue with lawyers dealing with construction to make the difficulties of managing engineering safety more apparent to all concerned.

Two solutions which have been proposed concern insurance and the greater use of 'approved' designers. In particular, concerning the law of limitation it has been proposed [8] that the whole of the risk and its insurance should pass to the owner of the structure five or six years after completion. The second solution, not independent of the first, is to make more use than is currently the case, of approved panels of engineers for particular categories of structures. Engineers on the panels will have specialist knowledge and skills with respect to structures in that category and of course they will be judged, in law, by the standards of other engineers on the list. Insurance premiums for the panel engineers should be lower if insurance companies perceive that risk of failure (particularly due to human factors) is significantly reduced.

Clearly there may be other solutions to these problems. It seems that engineers, safety analysts and lawyers must cooperate to discuss alternatives. In view of the inherent difficulties in producing high-quality estimates of failure likelihoods it may well be that safety analysts have a responsibility to take the initiative in setting up their discussions.

Conclusions

(1) A fundamental question for all engineers and particularly for safety analysts is, "Should society expect no failures?" The answer must be

no because such perfection cannot be reached and measures of risk are, by their nature, problematical.

(2) It is argued that engineering is a process of problem solving and decision-making using information with varying degrees of dependability. Dependability is necessary but not sufficient for truth.

(3) The notion of the inductive reliability of a hypothesis should be replaced by the notion of a responsibility to act on that hypothesis. The responsible engineer takes all reasonable precautions against being wrong and is careful, informed and above all self-critical. Judgements of reasonableness are made by the peer group whose judgements in turn ultimately depend on their ethics.

(4) The argument presented directs attention to the idea that measures of risk are not absolute measures but aids in the process of the management of knowledge to control risk.

(5) The recognition, study and consequent understanding of processes analogous to Kuhn's paradigms in the development of engineering, should ensure that the need for peer group approval does not lead to stagnation.

(6) Responsibility and reasonableness are central concepts to lawyers. In order to help the law to operate fairly and effectively, safety analysts may well have a responsibility to set up a dialogue between themselves, practising engineers and lawyers in order to resolve the current legal problems. Solutions may include different insurance arrangements or the greater use of panels of approved engineers for various categories of structures.

References

[1] Blockley, D. I. (1980). *The Nature of Structural Design and Safety*. Ellis Horwood, Chichester.

[2] Blockley, D. I. and Henderson, J. R. (1980). Structural failures and the growth of engineering knowledge. *Proc. Inst. Civ. Eng., Part I*, 68, 719–728.

[3] Lakatos, I. (1978). *Mathematics, Science and Epistemology — Philosophical Papers, No. 2*. Cambridge University Press, Cambridge.

[4] Cohen, L. J. (1977). *The Probable and the Provable*. Clarendon Press, Oxford.

[5] Settle, T. W. (1969). Scientists: Priests of pseudo-certainty or prophets of enquiry? *Sci. Forum*, 2, No. 3, 21–24.
[6] Thomas, N. P. G. (1984). *Professional Indemnity Claims*. The Architectural Press, London.
[7] Collingridge, D. (1980). *The Social Control of Technology*. The Open University Press, London.
[8] Ludlow, M. (1984). No winners in litigation. *New Civ. Eng.*, 6, 21–22.

Chapter 5

Reflective Practice in Engineering Design*

Abstract

Civil engineers are valued by society for their knowledge and skills which, though based on theory, are forged through experience. Indeed, the real world is a messy place of complex, interacting systems which time and time again demonstrate the inadequacies of engineering theory. The scientific approach to engineering education does not equip graduates effectively for professional practice. This chapter introduces the concept of 'reflective practice', explains how it can be formalized and suggests that it should be adopted alongside the traditional technical rationality approach to teaching engineering design. It involves analyzing past failures, using artificial intelligence and, above all, looking at the big picture.

Introduction: Professional Practice

The knowledge of a professional is based on theory but acquired through years of experience. So, what is the nature of this professional knowledge? The *reflective practitioner* [1] describes the mental processes of the professional as thinking in action and, when those processes are more

*This chapter was originally published in *Proc. Instn. Civ. Engrs, Civ. Eng*, 1995, 108, 160–168 (co-author: W. P. S. Dias).

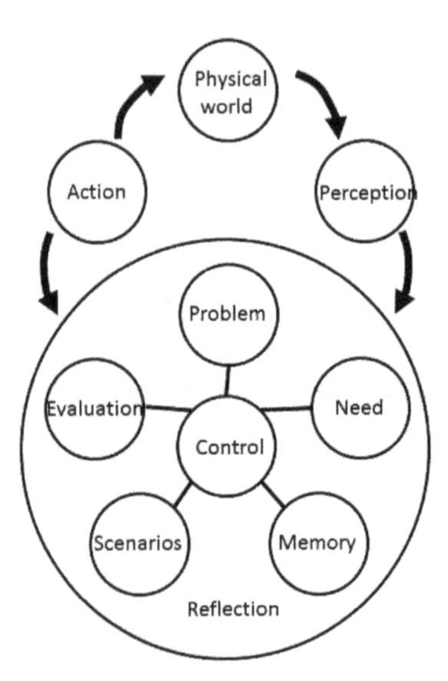

Figure 5.1. The RP loop.

deliberate, as reflection in action: hence, the term 'reflective practice' (RP), where there is a vital engagement by the practitioner with the actual situation at hand. These ideas have been developed [2] into an RP loop which consists of an interconnected set of *world–perception–reflection–action–world* loops (Figure 5.1). Reflection can be deep or shallow, depending on whether the action is carefully thought out or instinctive. The interaction leads to learning [3] which includes what has been called tacit knowledge [4]. It is as a result of this tacit knowledge that professionals often seem to know more than they can articulate.

The knowledge of professionals is normally held in great respect, but recently there has been a crisis in public confidence [1, 2]. There have been instances (for example, in the city) where professionals have misused their knowledge for personal gain; this is an issue about ethics. There have been accidents and failures — for example, at Ramsgate ferry in the UK [5], where a 20 m high walkway collapsed and six people died on 14 September 1994 and the collapse of the Sampoong department store in South Korea.[a] These events have suggested technical incompetence. One of the biggest problems for professionals is that the world is

[a] See http://en.wikipedia.org/wiki/Sampoong_Department_Store_collapse (Accessed on June 2019).

Table 5.1. TR and RP paradigms.

Paradigm	TR	RP
Characteristic		
Basis	Selective inattention	Reflective interaction
	Science	Systems
	(analytical)	(synthetic)
	(reductionist)	(holistic)
	(context independent and practitioner independent)	(context dependent and practitioner dependent)
Models		
Grounding	Truth	Dependability
Specification	Precision	Appropriateness
Improved by	Calibration against world	Comparison with world
Techniques	Mathematics	Artificial intelligence (AI)

increasingly complex — a 'mess', according to 'soft systems methodology in action' [6]. Infrastructure development, for example, involves environmental impact, end user satisfaction and disruption to human activity. Projects involve co-ordination between large numbers of organizations. A balance has to be achieved between quality (which includes safety) and cost. Complex human factors are important in minimizing the risk of failure.

Professionals sometimes find it difficult to handle these 'messes' because their education has not helped them to develop the necessary skills. Many engineering curricula are still dominated by engineering science in which well-defined physical phenomena are described — largely mathematically. Such knowledge is indispensable but covers only a narrow spectrum of the expertise of an engineer.

The quantitative approach to well-defined problems has arisen from technical rationality (TR) which governs almost all academic institutions. TR and RP are contrasted in Table 5.1.

Technical Rationality

TR is based on the scientific reductionist method where behaviour of the whole is explained from the behaviour of the parts. It is characterized by selective inattention to aspects of problem that cannot be formulated theoretically. It seeks to be practitioner independent or objective — so that results are not supposed to depend on the scientists. Abstract generalisations attempt to make it context independent. Problem solving is seen as the delivering of a product. Mathematics is the language that enables the formalization and generalization of TR. Many disciplines now seek respectability through mathematics. Quantification is the logical end result of the reductionist process.

A model is a representation of the world. Within TR the goal of the model is truth, i.e. correspondence is sought between the results of the model and the behaviour of the parts of the world being modelled. Precision and accuracy are therefore very important.

The TR approach has been enormously successful in advancing scientific knowledge over the past 300 years. At a different level, since the middle of the last century it has also helped in a limited way in management problems through operations research (OR). However, limits to mathematics and physics have arisen in the form of Gödel's theorems, Heisenberg's uncertainty principle and chaos theory. Also, many descriptions of scientific progress [7] reveal the extent to which it depends on social context. Furthermore, selective inattention is becoming inappropriate in today's increasingly complex and interdependent world.

In some cases, aspects that are difficult to cast into theoretical formulations (such as sociological, political and cultural considerations) may be the ones that most influence a solution — especially where engineering infrastructure projects are concerned. There is therefore a need for an alternative approach: one that emerges from professional practice itself.

Reflective Practice

RP is a possible alternative to TR. The essence of RP is reflective interaction based on a systems approach. The world is seen as consisting of a hierarchy of entities, each being a whole as well as a part, i.e. a 'holon' [8]. Each whole is greater than the sum of its parts in the sense that it has emergent properties. For example, the capacity of a human being to walk and talk is a result of the co-operation of the sub-holons of the body, such

Figure 5.2. RP views a structural frame as a system which is greater than the sum of its parts.

as the skeleton and nervous system, each of which on its own does not have these properties (Figure 5.2). Similarly, the corporate behaviour of a project team is undoubtedly influenced by the characteristics of its members, but the team also has a behaviour of its own, which emerges when the members are put together. The same is true of a structural frame consisting of columns and beams, which are made of the reinforcing steel and the surrounding concrete (Figure 5.3). The entities at each hierarchical level are both wholes and parts and the hierarchy is open-ended at both top and bottom. In that way a systems approach tends to be holistic, as opposed to reductionist. This holistic outlook makes the systems approach very appropriate for design.

RP is, by definition, dependent on both practitioner and context. Problem solving is seen as a process. The grounding (the basic concept on which something is based) of RP models is dependability, not truth. Correspondence with the world may be only approximate (as appropriate), but dependability is necessary for safety and serviceability. RP models may be graphically rich pictures (Figures 5.4 and 5.5). The emphasis

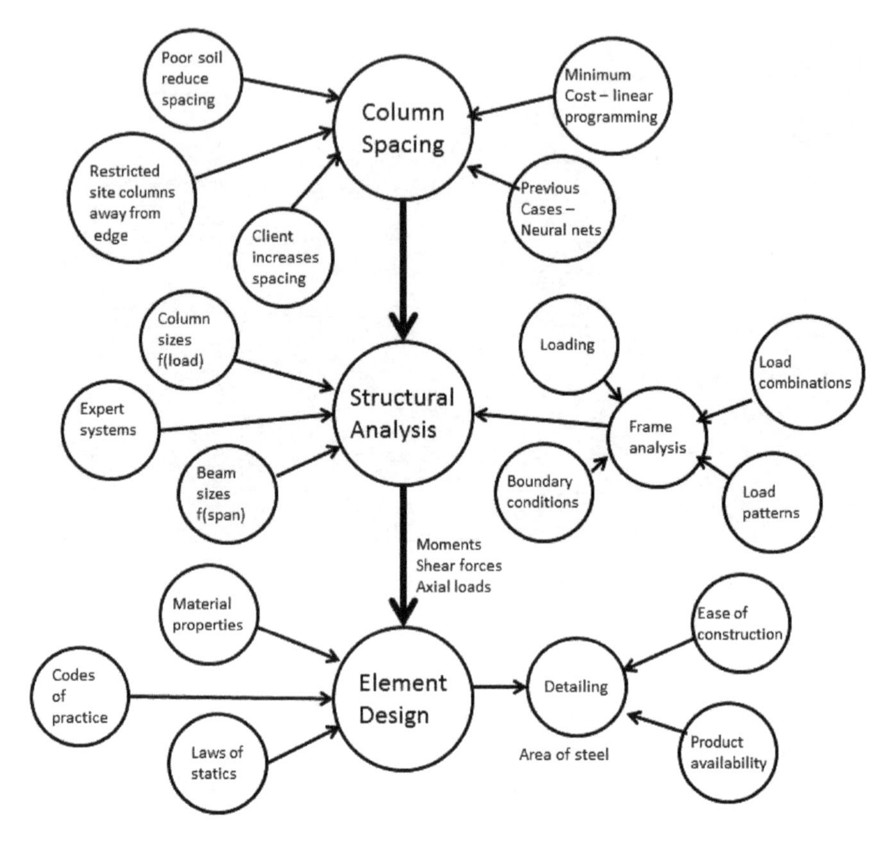

Figure 5.3. Rich picture for reinforced concrete frame design.

is not on calibration of the model for accuracy and truth but on continuous comparison of the model with the world, leading to greater understanding of the world and a more dependable model for good decision-making.

RP appears to be more relevant to professional practice than TR. Why then is it that RP has not found a place in academic institutions? One reason is that RP is a relatively new idea. The other — possibly more fundamental — reason is that RP lacks a formalism.

Formalizing RP

Practice often tends to be held in low academic esteem because of a lack of theoretical rigour. It is often seen merely as a collection of particular

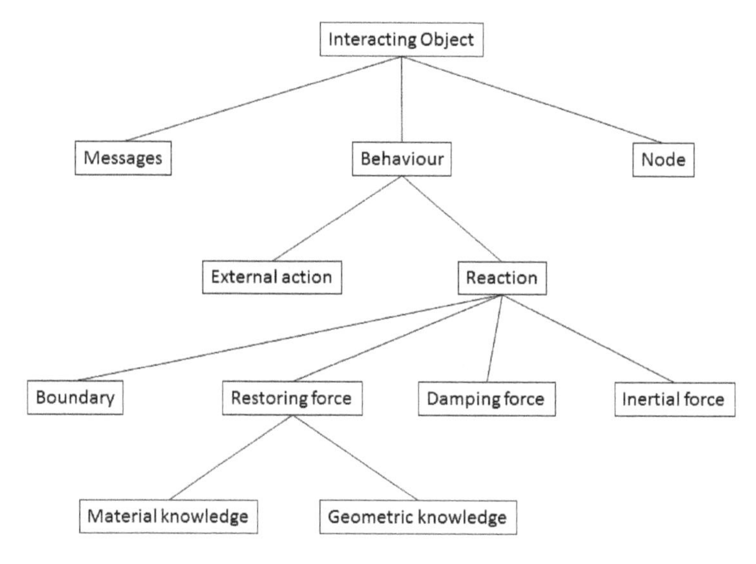

Figure 5.4. Structure of an interacting object.

activities. Formalism enables generalization and hence the development of an academic discipline.

Systems

There have been various attempts to define a systems language for disciplines as varied as biology and economics. A systems approach could provide the conceptual formalism for RP.

The following are the core concepts of a systems approach:

- Interconnectedness of hierarchically arranged concepts. A member of the hierarchy is seen as both a whole and a part, i.e. a holon [8]; this means that each whole is greater in significance than the sum of its parts. In other words, holons display emergent properties.
- Process loops with interaction and feedback, and the participation of the change agent in the process, leading to learning [3, 6].

OR, although called a systems approach chiefly because of its inclusion of interconnectedness, tended to be reductionist and strongly mathematically

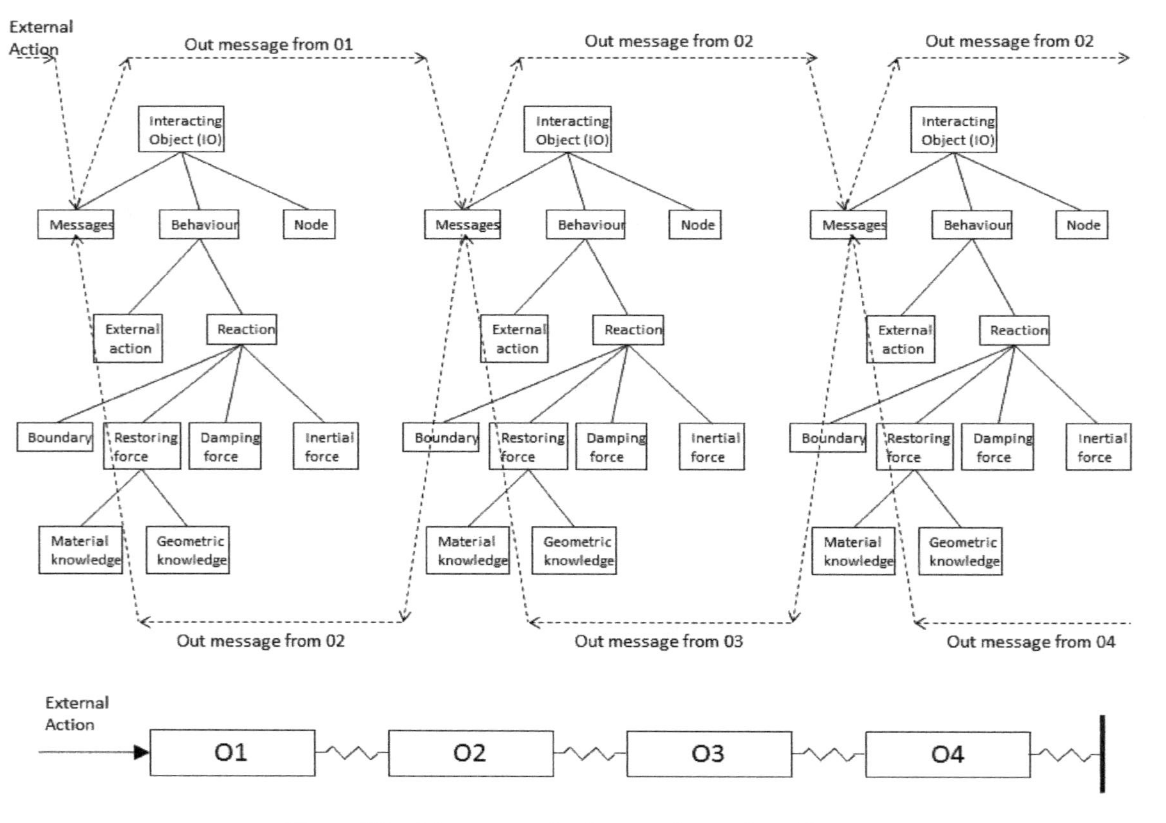

Figure 5.5. Message passing between interacting objects.

based. Systems approaches which are not strongly mathematical but rely rather on graphical representations, for example, are sometimes called 'soft' systems. This distinguishes them from the 'hard' systems approach of OR (see Chapter 8). One practical manifestation of a systems-based formalization is diagramming. Two graphical images are often used in systems — a pyramid (signifying hierarchy) and a loop (signifying interaction). A professional often uses diagrams to help the thinking through of a problem. Such diagrams or rich pictures might depict hierarchical structure, interaction and feedback. As the diagram is a qualitative representation of the problem, all aspects of it can be represented — not merely those that are amenable to theoretical formulation or quantification.

For example, in the design of a multi-storey building, a hierarchical decomposition diagram could reveal its component systems (structural, architectural, thermal, services and so on) and also indicate the influences that these systems have on each other (e.g. the type of ducting would influence the ceiling space). In addition, an external envelope diagram could reveal the socio-cultural context of a building, e.g. transport services, public amenities such as hospitals, shops and schools, and environmental quality.

Object-oriented programming [9] provides a computational language for systems. This is essentially a form of declarative programming with the emphasis on the description of a given situation in data structures called objects. Procedures are more important in TR, and hence procedural programming languages such as Fortran are used. In object-oriented programming, data and procedures are contained within the objects, which interact by sending messages to and receiving messages from other objects. Objects can be structured hierarchically by being declared as descendants of others so that they inherit attributes. For example, a steel beam object inherits all attributes of a beam object, in addition to having special attributes of its own. Objects can also function as aggregates of other objects, by having lower-level objects declared as attributes of a higher-level object.

For example, a frame object can have beam and column objects as its attributes. In this way, object-oriented programming supports both generalization/specialization and aggregation/decomposition hierarchies.

The interacting objects process model [10] is a practical manifestation of the systems approach to basic mechanics (Figure 5.5). Laws of physical behaviour are programmed into interacting software objects which hold local information. The processes are massively parallel and potentially enormously powerful.

Within academic engineering curricula these systems concepts could be taught explicitly. However, just as the scientific method is communicated implicitly rather than explicitly, a systems approach to problems should be encouraged in the teaching of all subjects.

Artificial Intelligence

The fundamental change in computing which gave rise to AI was the development of the ability to process symbols as opposed to numbers. The initial goals were very general and ambitious, for example, to mimic human intelligence. Later, expert systems were written using rules to capture and store large amounts of expert knowledge of specific domains [11]. An example which combines expert opinion and measured data is the SIPIT system for interpreting pile integrity tests (Figure 5.6) [12].

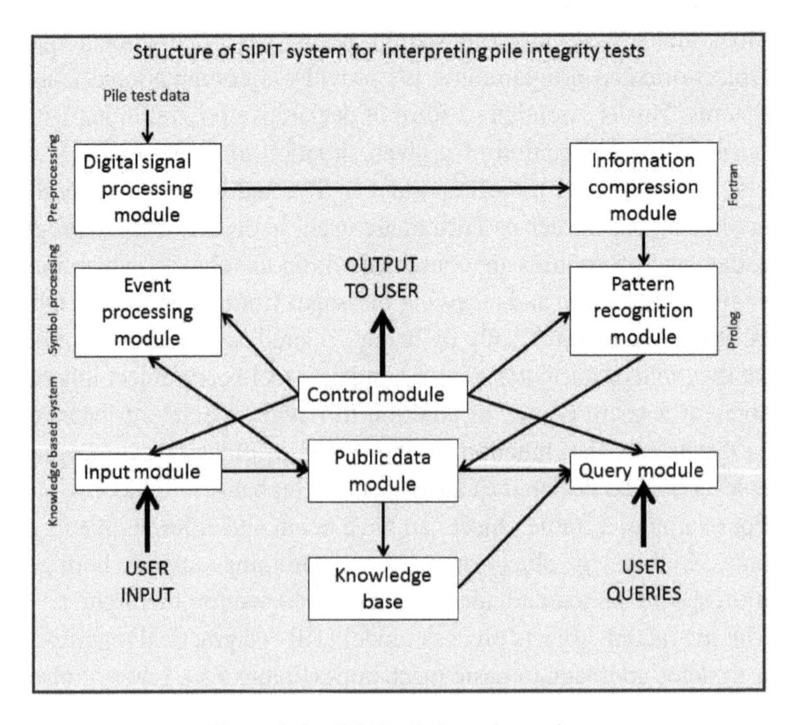

Figure 5.6. SIPIT pile integrity testing.

Recently the connectionist view of knowledge has emerged. This does not force the expert to codify this knowledge in the form of rules. Rather, experience is expressed in the form of case histories that are used to train the computer. Techniques are being developed for the computer to discover patterns of which even the expert may have been unaware. Artificial neural networks are probably the best examples of this approach.

AI can be considered as a generalizing language or tool for RP because the same AI techniques (such as expert systems and neural networks) can be used in a range of professional practices. The goals of AI are concerned with human intelligence, experience and even artistry, which is what RP is about too. Obviously much more research is required.

In summary, RP can be formalized at the conceptual level through a systems approach and on the technical level through AI (i.e. symbol processing).

Use of RP and TR in Engineering Design

TR and RP are distinct approaches to engineering design. A simple example in structural engineering design will be used to illustrate the approaches of RP and TR. Consider the design of a reinforced concrete building frame.

The discussion here is summarized in Table 5.2. An RP approach would involve, for example, an appreciation of the client's needs and requirements, the site conditions (especially the soil conditions) and the building of project teams. These factors may be represented initially in a rich picture (Figure 5.3). In addition, an RP approach would seek to draw on any previous experience of the design of such frames. In the future a more formalized approach might involve AI. For example, a neural net might be trained on case histories in order to discover some relationship between, say, column spacing and the parameters that affect it (such as bay size, soil conditions and building function). After such training, values or symbols representing the parameters of the current design can be fed into the trained network to obtain a recommendation regarding column spacing for that design.

The designer may then wish to focus on a cost minimization exercise to determine the number and/or spacing of columns; this is work in a TR

Table 5.2. Example of the use of RP and TR in engineering design.

	RP		TR		
Design phase	Reflective interaction	AI techniques	Focus of attention	Mathematical techniques	Relationship of RP to TR
Choice of layout	Clients requirements, site, soil conditions, past experience	Neural network	Cost minimization	Linear programming	Contextual
Structural analysis	Initial sizing, idealization	Expert systems	Structural response	Frame analysis	Precondition
Element design	Ease of construction, product availability		Cross sectional equilibrium	Laws of statics	Constraint

mode. Cost is only one of the considerations involved (note the selective inattention of the TR paradigm), especially because structural cost may not be a very high proportion of total cost. Nevertheless, a linear programming technique could be adopted — perhaps to fine-tune the decisions that were taken based on the RP considerations.

A next step in design would be the structural analysis of the frame. The TR approach will focus on the actual structural response that could be obtained by performing a frame analysis. However, before this the beams and columns have to be sized and the frame has to be idealized (e.g. as rigid joints and fixed bases). Sizing and idealization are normally based on heuristic knowledge and would therefore be part of the RP paradigm. In fact, rules such as found in a design manual [13] regarding sizing could be fed into a knowledge-based system that could be used to perform the initial sizing. Such a system would contain rules such as 'if a beam is of type continuous and carries a udl (uniformly distributed load) of value medium, then the beam depth is span/12'. When the system is fed with facts concerning the current design, the relevant rules will be triggered in order to suggest an element size.

One of the final steps is element design, where beam and column reinforcement is determined. The TR approach will focus on using the laws of statics to determine the quantities of reinforcement, so that the cross-sectional equilibrium of elements is satisfied. However, an RP approach will constrain the specifications to satisfy the reinforcement sizes available in the market. In addition, an RP approach will limit or rationalize the different sizes and spacings of reinforcement that are specified, in order to facilitate construction.

In the three steps of the design process described, three ways can be identified in which RP encompasses TR. In the choice of layout, the RP approach provided the context for TR. In the structural analysis, the RP approach was a precondition for TR. In element design, the RP approach was a constraint on TR. These relationships can be depicted graphically as in the last column of Table 5.2. The example given here shows therefore that RP and TR are not mutually exclusive approaches where engineering design is concerned: rather they are complementary, with RP having a broader focus than TR.

In current design practice the RP approach is the engineering approach without the corresponding AI techniques. However, as RP becomes more recognized some formalized AI techniques will also be seen as a natural development.

Can RP help to produce genuinely innovative design? Although there are attempts to use AI techniques for this purpose [14] it is difficult to formalize innovation at this technical level and this is a subject for much further research. However, RP formalisms used at the conceptual systems level to develop rich pictures and process loops can foster innovation. Creativity in engineering is very often results from capitalizing on the particulars of a given design through reflective interaction.

Implications for Engineering Practice

So what are the implications of RP for engineering education and practice? Something that springs directly from the idea of reflection is the importance of case histories. These are very important in the medical and legal professions and should be given more formal recognition in engineering. Engineering students as well as practising engineers could upgrade their knowledge vastly by learning from case histories of design and construction and of failure.

Reflection on failures will result in improved design and construction [16]. Event sequence diagrams, for example, can represent the essential preconditions to failure (Figure 5.7). Where engineering design

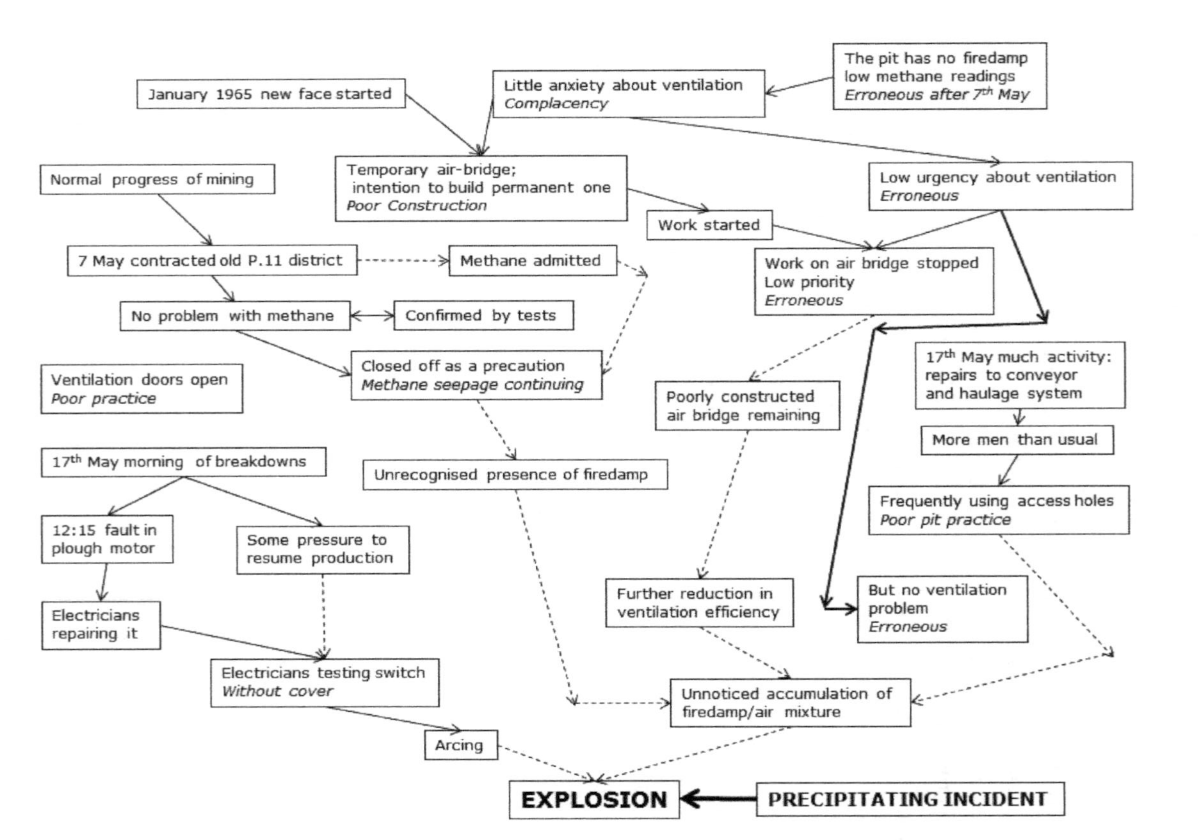

Figure 5.7. Event sequence diagram for a colliery explosion [15].

companies are concerned, a concerted effort should be made to document and/or computerize their experience on previous projects.

The importance of rich pictures in the systems approach should also be stressed. The craft-based tradition of engineering drawing tended to be very technical and concentrated on working drawings, reinforcement details, circuit diagrams and so on. A more conceptual type of drawing that has emerged with the computer age is that of the flowchart. Flowcharts also tend to capture mainly those aspects of the problem that are amenable to theoretical procedures; hence, they could be described as characteristic of TR. A rich picture, however, has no limits on what it can include. It is also subjective and practitioner dependent; this means that the experience of the practitioner can be reflected in the rich picture. All three types of graphics — rich pictures, flowcharts/networks and working drawings have their place in engineering practice. However, the rich picture is perhaps the least understood and appreciated, although it is often used intuitively by practising engineers to discuss ideas. These graphical techniques could be developed further for more formal use in engineering undergraduate design curricula.

AI techniques could also be given more prominence both in education and practice. Practitioners could be encouraged to make use of many systems that are fast becoming commercially available. They could also use AI techniques to capture their own knowledge, or the collective knowledge of their companies. Some of the principles of AI and object-oriented programming should also be introduced into engineering curricula.

Conclusions

(1) Academic institutions are currently dominated by TR which is based on the scientific method and characterized by processes of selective inattention. This results in a tension between professional practice and its theoretical base. RP, which is a systems approach to problem solving involving reflective interaction, is a much better framework for understanding and developing a theoretical base for professional practice.

(2) In engineering design, RP and TR approaches are complementary. The focus of RP is wide, whereas the focus of TR is narrow. RP may be a context for, a constraint on and a precondition for TR.

(3) RP can be formalized conceptually using systems concepts and technically through AI. Such formalization will help RP to gain more academic respectability and become incorporated in engineering curricula.
(4) RP can be encouraged in engineering education and practice by emphasis on the importance of case histories and by the use of rich pictures and AI techniques.

Acknowledgements

The authors were able to collaborate with each other on this chapter as a result of a Commonwealth Fellowship awarded to W. Dias at the University of Bristol, while he was on sabbatical leave from the University of Moratuwa, Sri Lanka.

References

[1] Schon, D. A. (1983). *The Reflective Practitioner: How Professionals Think in Action*. Temple Smith, London.
[2] Blockley, D. I. (1992). Engineering from reflective practice. *Res. Engng. Des.*, 4, 13–22.
[3] Senge, P. M. (1990). *The Fifth Discipline — The Art and Practice of the Learning Organisation*. Century Business, New York.
[4] Polanyi, M. (1967). *The Tacit Dimension*. Routledge and Kegan Paul, London.
[5] Chapman, J. C. (1998). Collapse of the Ramsgate walkway. *J. Struct. Eng.*, 76, No. 1, 1–100.
[6] Checkland, P. and Scholes, J. (1990). *Soft Systems Methodology in Action*. Wiley, Chichester.
[7] Kuhn, T. S. (1962). *The Structure of Scientific Revolutions*. University of Chicago Press, Chicago.
[8] Koestler, A. (1967). *The Ghost in the Machine*. Picador, London.
[9] Stefik, M. and Bobrow, D. G. (1986). Object-oriented programming: Themes and variations. *Artif. Intell. Mag.*, 6, No. 4, 40–62.
[10] Chandra, S. *et al.* (1992). An interacting objects process model. *Comput. Systems Engng.*, 3, No. 6, 661–670.

[11] Adeli, H. (ed.) (1988). *Expert Systems in Construction and Structural Engineering*. Chapman & Hall, London and New York.
[12] Comerford, J. B. *et al.* (1991). The interpretation of pile integrity tests through intelligent knowledge-based systems. *Proc. Instn. Civ. Engrs.*, Part 1, 90, 189–204.
[13] Institution of Structural Engineers (1985). *Manual for the Design of Reinforced Concrete Building Structures.* IStructE, London.
[14] Coyner, D. *et al.* (1990). *Knowledge-Based Design Systems.* Addison-Wesley, Reading, Mass.
[15] Turner, B. A. (1978). *Man-Made Disasters*. Wykeham, London.
[16] Blockley, D. I. (1980). *The Nature of Structural Design and Safety.* Ellis Horwood, Chichester.

Chapter 6

Thinking Outside of the Box with Phil's Eight New Maxims*

Abstract

This chapter is a slightly edited version of my Presidential Address to the Institution of Structural Engineers in 2001. At the time, the UK professional engineering institutions were being accused by some as being old fashioned and out of date. I said that, as engineers, we lack a human face to the public. The public see us often as dealing with prescribed things — applying science with little room for judgement and little room for human values. I said that we needed new perspectives and skills in order to see ourselves as others see us. We need to avoid being technically narrow and narrowly technical. I suggested that we should understand and nurture practical rigour and practical intelligence much more clearly. As we all face the new uncertainties of the 21st century there is undoubtedly a need for a 'common voice' from all engineering organizations with some calling for them to amalgamate into one huge organization. I said that the value of belonging to a specialized smaller body like the Institution of Structural Engineers for me was huge and I gave six examples. I quoted research that shows that belonging to any social group is seriously good for your health. I argued that the common

*This chapter is an abridged version of my Presidential Address to a meeting of the Institution of Structural Engineers on Thursday 4 October 2001 and was originally published in *Struct. Eng.*, 2001, 79, No. 20.

engineering voice should emerge from a 'systems thinking' approach across all engineering organizations so that we all articulate coherent and clear messages to our various constituencies.

Introduction: My Objectives

It is an honour and a privilege to have been elected President of this great Institution of ours — the Institution of Structural Engineers (IStructE). I am acutely aware that as President I am in illustrious company, including that of my famous Bristol predecessor Sir Alfred Pugsley [1]. I will do my very best to serve, represent and lead the Institution.

In this address, I want to share some insights — Phil's new Maxims — and:

(1) articulate the value of the Institution;
(2) argue for new perspectives and skills;
(3) understand practice in theory better;
(4) look forward to some new research futures;
(5) suggest some IStructE messages;
(6) draw some conclusions.

Articulating the Value of IStructE More Clearly

Where Am I Coming from?

I am an 'academic engineer', so what does that phrase conjure up in the public mind? In every-day language 'academic' means 'irrelevant' and 'engineer' means 'mechanic' — so to many people I am an 'irrelevant mechanic'!

Academics are often accused of having their 'heads in the clouds'. But many academics are also involved in intensely practical things. For example, running a faculty with a £20 million turnover, in a modern university where money is earned not granted, as I have done, is a challenging practical responsibility these days. Acting as a structural consultant and as a non-executive director to a plc also helps me to keep my feet on the ground.

My research has been driven by intensely practical motives but one can be forgiven, if you have read some of my writings, for thinking that my feet are some way up in the air. The norm in research is to develop a theory and see if it works in practice. I have also become intrigued by the opposite question — looking at what works in practice and wondering if it can work in theory.

This is because I believe there is no adequate theory of practice. I will need to explain. I feel that, intellectually, we engineers have allowed ourselves to be browbeaten into feeling we are *ad hoc* and non-rigorous because we cannot and do not attempt to provide truth and certainty. Hence, the frequently heard 'get out clause': 'I am only a simple practical engineer'. Science (the pursuit of precise truth) is only part of engineering. I believe that practical decision-making (and specifically engineering) is vastly under-rated as an intellectual activity.

You may be thinking — well, so what? I believe that a theory of practice may provide some (not all) of the keys by which our profession may unlock the recognition it deserves. One thing is for sure: we have to earn recognition and status by our own actions — no one will present it to us on a plate! We have to stop apologizing for being practical and assert our place in intellectual life.

We Have Been under Attack

The professional Institutions have been under some pressure. They have been accused of being old fashioned and out of date. For example, Professor Alexander Kennaway wrote in the *New Scientist* as long ago as 1989: 'In my view the Institutions have outlived their usefulness. Surely the way forward is for Institutions to dissolve themselves, subsuming their residual functions of status seeking, the award of titles and the overview of professional practice experience into the Engineering Council'. More recently in a report 'Rethinking the Institutions' to ICE in 2000 [2], it was stated that there is a widely held view that the Institutions are out of touch.

There are a number of engineers who think we should update the profession by amalgamating into one large UK Institution of Engineers.

I believe that the latter view springs from addressing the question 'how' before the question 'why'. I suspect it stems from a desire to have a 'common voice' for engineers — something clearly we do need. However, it is not necessarily a need that is best served by making all engineers the same — no more than it would make sense to say that all of the cells in your body should be the same because they are part of you. We need to celebrate inward diversity and look outwards to find this 'common voice'.

I think a move to a single Institution (even if it were to contain specialist colleges) will be impractical for a very long time, and may be always, if a requirement is that it should enjoy the sort of 'bottom up' support that our Institution has. Hardly any of us will commit to a top-down-managed single Institution. People have to be proud of their Institution to want to give their support. At the Structurals many members commit themselves to a lot of time and effort. The success of our Institution is a tremendous tribute to every member who is proud to be actively involved in our many and various activities at HQ in London and in our Branches and Divisions all over the world in 105 different countries.

It is clear that now is a time of great change in our industry and in our profession. We structural engineers must embrace change if we are to get the things we want.

I Get Good Value from IStructE

As Brian Clancy said in his 1996 Presidential address, 'we are the Institution — the Institution is us'. We make it what it is. Therefore, I would like to start by telling you what the IStructE means to me.

These are the six major reasons why I, personally, value the Institution:

(1) Having the qualification. The Part 3 exam is a tough test — I am proud of having passed it;
(2) Meeting and learning from other structural engineers, especially locally. Nowadays some call it networking (a horrible word that makes it sound exploitative) but really it is just making friends and business contacts and, through that, influencing people inadvertently;
(3) Being aware of change in our business, of different ways of doing things and of what can be done;

(4) Intending that, through being aware, some of the good practice will rub off on me;
(5) Exercising a legal duty of care to my job;
(6) Finally, making a contribution.

As I will discuss later, our knowledge and skills grow by critical discussion. This is not just a superficial observation but a deep philosophical point that is at the 'heart' of what the Institution is all about. Of course, with respect to the last point above, 'making a contribution', the old adage is that 'You only get out of things what you put in'. All I would say to those who use the Institution only as a 'meal ticket' is that, in my direct experience, there is so much more to be had. I firmly believe that the Institution exists to help members to add value to themselves. Adding value is a 'buzz' phrase that simply means 'making something worth more'. When you pass the Part 3 exam you have added value to yourself. I believe that to add further value to yourself you first add value to others. What better way than by taking an active role in the affairs of the Institution and at the same time demonstrating clearly your duty of care?

The Institution is Seriously Good for Your Health

Putnam [3] has demonstrated quite vividly how being involved actively in an organization such as our Institution can make you healthier and happier. He writes: 'If you smoke and belong to no groups, it is a toss-up statistically whether you should stop smoking or start joining'. He came to this somewhat startling conclusion by calculating, from the USA national census data and other sources, an index of something that the sociologists call social capital. He did it for each of the states in the USA. Social capital is the extent to which people connect or are 'joiners' in social activities with the norms of reciprocity and trustworthiness. He then correlated these measures with other data such as mortality rates and various other measures of well-being (Figure 6.1 where the points are the initials of US States). Of course, the results must be treated cautiously — they are statistical, not causal, results and explanations are not easy. Putnam discusses them at length in his book.

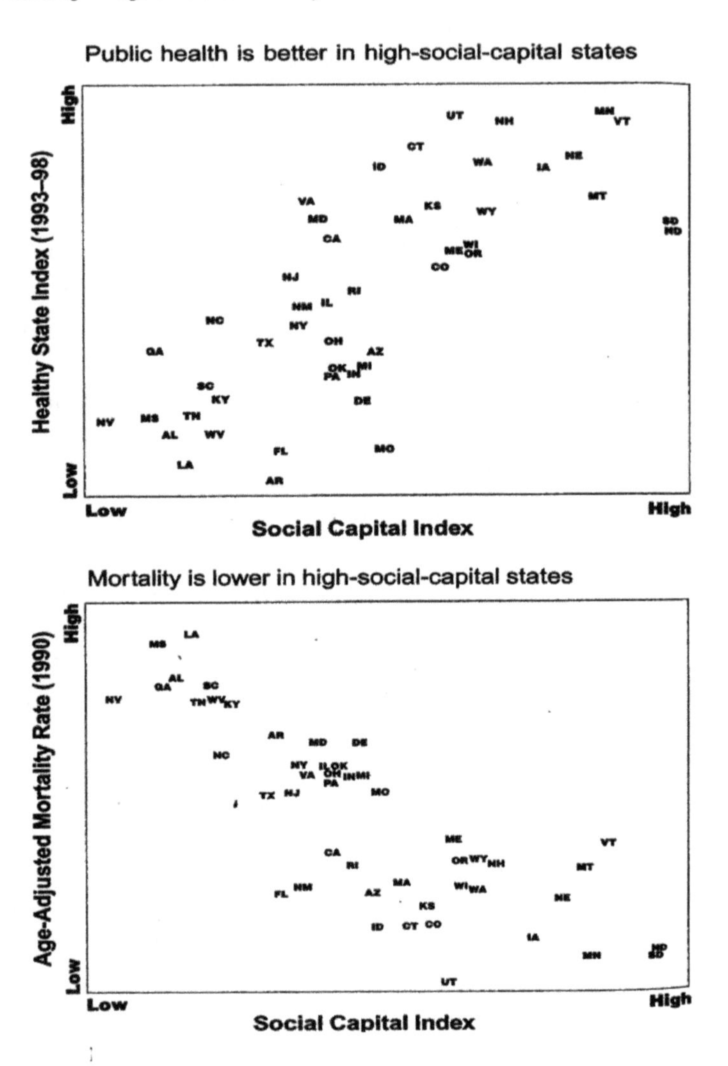

Figure 6.1. Health is better in high social capital states of the USA.
Source: Reproduced with the permission of Professor Robert Putnam.

One of the important messages of his work that has implications for our Institution is that social capital in the USA has been declining since the 1960s. Figure 6.2 shows membership of voluntary American professional societies. Notice the dramatic drop-off in the 1990s which is found in many different aspects of the USA life. As far as I can determine, no one has performed an equivalent study in the UK or elsewhere in the

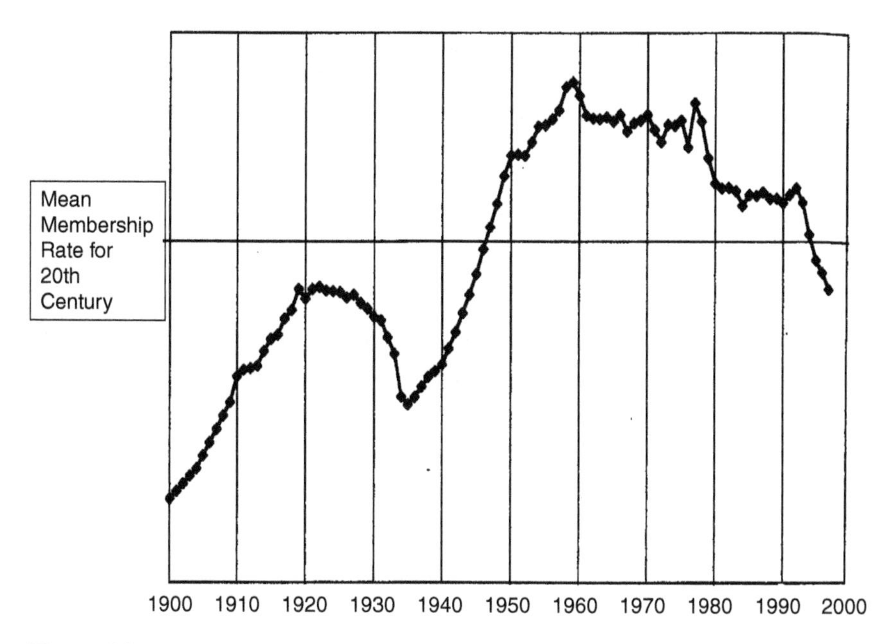

Figure 6.2. Average membership rate in eight US national professional associations from 1900 to 1997.

world, but where the USA goes many other countries follow. It is therefore quite possible that we are seeing a dramatic decline in social capital also in the UK. Attendance at Institution meetings has always been a concern but the Institution still enjoys a lot of support from members. Those from the 'silent' majority of non-active members should think about Putnam's results — the Institution could seriously improve your health as well as keep you up to date with best practice!

1980 Was an Important Year for Me

In 1980, I was honoured to be elected as Chairman of Western Counties, in the same year that Michael Horne was the fifth academic President. Michael's example was influential. I admired the way in which his scientific work underpinned his engineering. I have tried to follow this by digging deep into some of the issues I want to share with you tonight.

By 1980 I had developed a deep concern for the divide between theory and practice in structural safety and this was the subject of my first book, published in 1980. My Chairman's address was published in 1983 in *The Structural Engineer* under the title 'Phil's Eight Maxims' [4]. You will now see the reason for the sub-title of tonight's address.

There is a Need for New Perspectives and Skills

I Chose My Title on a Train

I chose the title of this address after a recent train journey. I was actually pondering a title when I overheard the following snippet of a conversation between two fellow passengers: 'We wanted someone who could think out of the box so we didn't want an engineer'.

I bristled inside when I heard that. However, on reflection I had to accept that the cliché may be true for some of us all of the time and for many of us some of the time.

We Need to Break Out

Our profession continues to do a great job with many impressive structures. Despite this, engineers do have the image of being rather narrowly technical and technically narrow. Whether we believe that to be the case or not, I think that we have to recognize and deal with the perception.

As I said earlier, I have come to the conclusion that engineering is something that works in practice but not in theory. Our theory is very partial — it is physical science theory that addresses only one aspect of our work and this leads us to look at the world in a very special way. One is reminded of another adage, 'if you only have a hammer then everything seems like a nail'. Why is it that:

- we are too often not too hot on people skills compared to other professions? For example, perhaps some of us could improve the attention we give to our clients;
- we often do not understand our clients' business too well? We think adding value is only about solving some technical problem, when the client is rather more concerned about the way the asset will improve his/her business;

- construction is one of the few industries where in-service measurement (i.e. the monitoring of strains, deflections, etc. in critical locations) is seen as a weakness and not a strength?
- even very able engineers sometimes feel it necessary to apologize for themselves and say, when faced with anything remotely outside their field, 'I'm only an engineer'?

We must set about sorting out together why we are under-appreciated. We do need to balance the necessary technical detail with a better understanding of the value we bring to the client and to society. Each of us must ask ourselves in specific situations: Precisely how are we adding value here? It is very easy to be adding value that goes entirely unnoticed. We need better skills in getting involved in wider issues and communicating the value we add — without compromising our technical contribution.

Some see change as a problem; we need to see it as an opportunity — often lurking in disguise. I believe that to get started on this we must get better at being able to see ourselves as others see us. This is the first of Phil's new Maxims:

Phil's Maxim No. 1: 'Look inside out and outside in'

Egan [5] asked us to rethink construction. I have written with Godfrey [6] that to do this we must think differently. We have to stand back and re-evaluate what we know and do, i.e. to 'get out of our box' and look at ourselves from the outside in. We must try to understand how others see us and realize our interdependence (Figure 6.3).

We Need to Understand Practice in Theory Better

As engineers, whenever we make a decision we must be able to justify what we have done. To understand this more deeply, we need to think about:

- how we know what we know;
- how we depend on theory for making decisions;
- 'wicked' problems;
- our values;

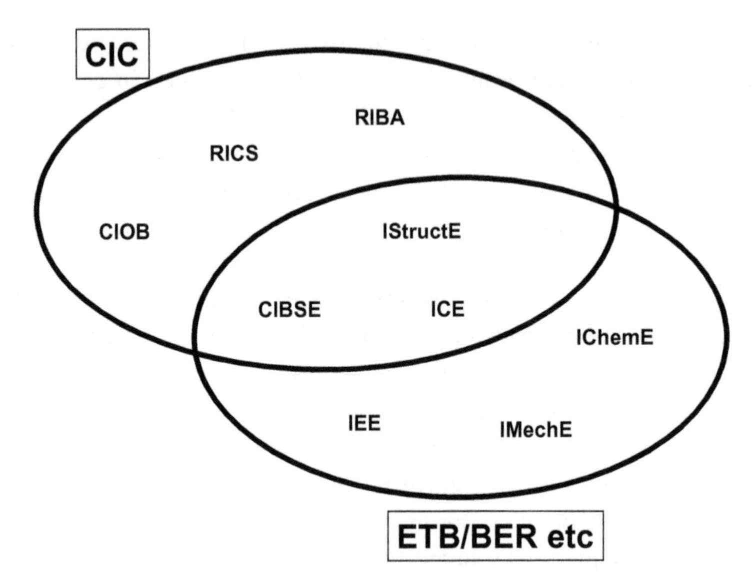

Figure 6.3. Construction and engineering clusters.

- how to justify the way we manage uncertainty and risk;
- practical intelligence and practical rigour.

Phil was Me — Logos

Perhaps I had better admit that Phil was me. I was asked after my Chairman's talk in 1980: 'Why did you choose the name Phil?' The answer was, rather on a whim, but Phil is short for Philo. Philo of Alexandria wrote in about 40AD about the 'Logos' — the totality of all ideas including the mind existing as patterns of things and causes of things. He addressed the problem of universals and particulars. *Logos* (in St. John's Gospel the 'Word' — 'In the beginning was the Word', John Ch 1, v1) captured the idea of the mind of God but also the intelligible world including the laws of nature and correct reasoning or logic.

Let us Clarify the Subjective and the Objective

I believe that the best way to think about the *Logos* is to use Figure 6.4 and Popper's Three Worlds [7, 8]. But what has this *Logos* got to do with

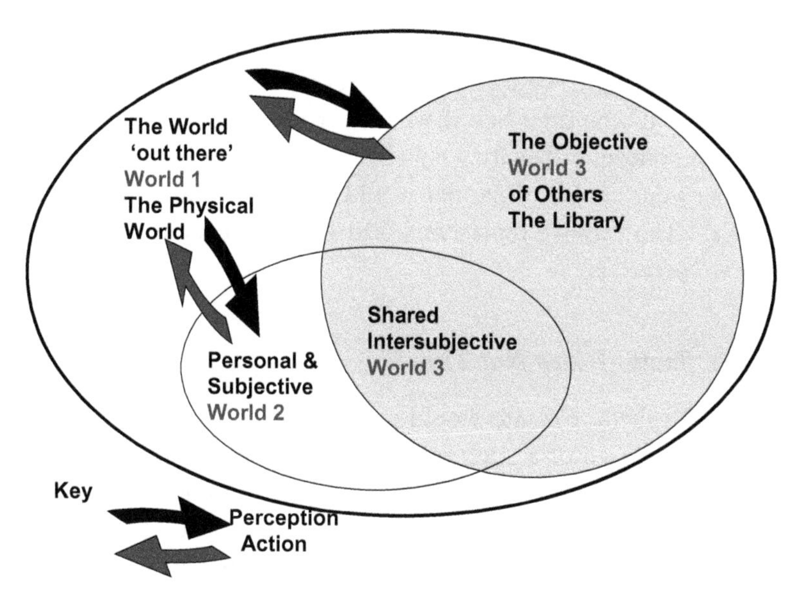

Figure 6.4. Logos: subjective/objective.

anything practical and structural engineering in particular? We have to make decisions that work and we have to be able to justify our decisions to others. Logic is about strict formal rationality — deducing conclusions from universals with total rigour.

It is part of that which we agree is objective, i.e. knowledge that is outside any one of us. Karl Popper called this his world of objective knowledge or *world 3*. Think of it as all of the libraries in the world and the Internet — but it contains fact and fiction. It contains all of the knowledge that we share, whether true or false. It includes unicorns, Sherlock Holmes, properties of steel and concrete, and concepts of force, stress and displacement. So, in this world of shared knowledge what is true and what is not? Philosophers continue to argue about this but we have to decide what we can depend on and use it to decide how to act.

All of us have our own private worlds of subjective, emotional and spiritual experience that Popper called his *world 2*. For example, it is difficult for me to tell you about the pain in my stomach. The physical world 'out there', outside any one of us, is what Popper called it his *world 1*. We can only experience world 1 through our world 2 senses. World 1 contains the structures we build. Common sense tells us that world 1 is there in front of us, we see it, we touch it, we feel it — it is obvious. However, imagine what it would be like if we saw world 1 with eyes sensitive to X-rays. Our concept of beauty would be quite different! You might reflect on whether your choice of partner would have been different if you had X-ray eyes! The way we represent world 1 is not absolute — it depends on how we perceive.

Facts are Facts — Are Not They?

We deal with facts. No one would seriously dispute that $2 + 2 = 4$ or that if a man lets go of a stick in his hand then it will fall to the floor. But such strict truth is elusive. In brief, we have to describe a context for any statement we make and then that context needs a context, and so on. In philosophical terms, we need a meta-language or a meta-theory. Let me explain. Firstly, we could write $2 + 2 = 11$ rather than $2 + 2 = 4$ and be correct. The reason is that the first is to base 3 and the second to base 10. Number theory is part of mathematics and the whole edifice depends upon set theory and formally defined axioms to define a completely rigorous language with no room for inconsistent conclusions.

Unfortunately, Gödel showed in the 1930s that any formal language (i.e. various mathematical theories with various axioms) rich enough to contain arithmetic contains inconsistencies in that there are theorems that are true but cannot be proved and some that can be proved but are not true. This means that the very basis of what we depend on absolutely — numbers — has absolutely fundamental flaws, *if you require absolutes*. Thus, the context of $2 + 2 = 4$ has a context and that context has a context and so on forever in an infinite regress. Likewise, if a man lets go of a stick in a spacecraft it will float. If he does it underwater it will rise. What happens when a man lets go of a stick depends on context.

Models are Partial

Theories and models are representations of aspects of the world. A physical structure is a 'hard system' — action and reaction — which has a

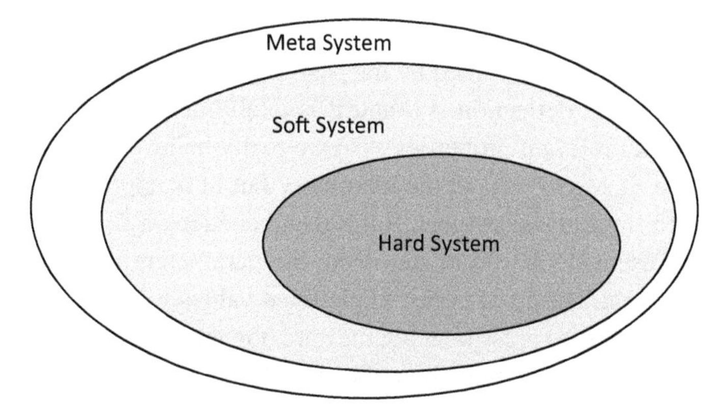

Figure 6.5. Structures in context.

context. It is embedded in a 'soft system' of people and organizations — action, reaction and intention. In turn that soft system is embedded in a meta-system. This contains everything else (hard and soft) that is outside of our concern and that we do not question (such as arithmetic is to base 10 and structures are on the earth's surface so sticks fall to the ground) (Figure 6.5).

Models:

- are incomplete since by definition not everything can be included;
- depend on what we are interested in and value, i.e. on that to which we give worth;
- are defined by our purpose;
- are dependable in a context and may break down outside that context;
- are essential to make decisions.

If you write down an expression for the deflection of a beam, or the bending moment in the beam, you say nothing about the cost of the beam or its beauty or its buildability, etc. — although they are not independent. If you calculate a reliability based on that model then it is a partial reliability which can be dangerous in the wrong hands.

We are taught deterministic theory. The word deterministic means that the future is totally determined by the past. If you put a load W on a beam of span l then the deflection is d and it is totally determined by W and l (with the other relevant properties). Do we really believe that? No — it is just a guide. It works well in the laboratory but only approximately with actual beams in actual structures. But is there a better way? Many theorists say yes — use a stochastic model. Here the parameters W, l are random variables so that instead of being single fixed values we use a probability density function to express how big they are. On the face of it this is much better since we are uncertain about the size of the load and the exact span. But this is only according to the model — there are many other uncertainties, too, some within the model and some (most) totally outside of the model.

So how do we know if these models are any good? We know, said Popper, by testing in world 1 (the physical world) and world 3 (objective knowledge) and it is this ability to test statements that demarcates scientific statements from non-scientific ones. It is a process of trial and error, of conjecture and refutation in critical discussion.

The net result of all of this is that we see the world, we decide and we act through our theoretical understanding. Theory is a pair of spectacles through which we view everything around us.

The Wicked Problems are Showing

How familiar does this sound to you?

- This problem is like sorting out a bowl of tangled spaghetti that seems to get more tangled the more you try to sort it out.
- I cannot get anything done these days — I have got another meeting to go to.
- Everything seems to take twice as long these days.
- The goal posts are constantly moving.
- There are so many people involved; the whole thing is a social process.
- You do not seem to understand the problem until you have a solution.

Some people call these 'wicked' problems as a contrast with 'tame'. Tame problems are linear and logical and we can sort them out step by step. Engineering science works with tame problems and we have achieved much this way. So we must not undervalue tame problem solving. However, if we use tame methods on wicked problems we will not get very far. Interestingly, non-linear methods of structural analysis are leading to problems with wicked characteristics such as 'chaos' and severe limits to predictability. Solving wicked problems is a 'messy' social process in which engineering science is a necessary but not sufficient part. I believe that we can navigate our way through the morass by focusing on some simple and fundamental questions. The first is always to start by asking the question WHY.

Phil's Maxim No. 2: 'Ask What Do They Want? What Do They Really Really Want? Then Give it to Them with Brass Knobs on'

I believe that until we are prepared to contribute to the wider purpose within a situation we will continue to be marginalized as technical experts. That is not to say we act in areas outside our competence as technical specialists but rather that we recognize that as team players we have a role in finding win–wins for everyone involved. The answers to why questions drive processes, they are the potential that drives the flow of change, the voltage that drives the current. The answers to why questions define the values and hence the quality of a project. The answers to how questions enable that flow. I characterize this symbolically as

why = *how* (*what, who, where, when*),

i.e. the *why* issues drive the way the how methods operate on the *what, who, where, when* parameters which describe the flow of change within a process.

For example, for the process 'Taking a journey' the reason *why* is the need to get to your destination, the *how* is the transformation function, i.e. by driving your car, and the *what*, *who*, *where*, *when* are the parameters that describe the car, the people, the place and the time.

The answers to *why* questions determine the appropriateness of a model, the values people hold and hence the meaning of quality.

Values are the Worth We Give

Making decisions means making choices and demonstrating a preference. We prefer those things that we value. In other words, a value is the 'worth' we give to something. Excellence is the state of having the highest value. Quality is not only fitness for purpose but it also is a degree of excellence and sometimes these are seen as different. Actually, they are not if you are absolutely clear about the values that define purpose.

Phil's Maxim No. 3: 'Know Your Own Values'

Know yourself and know what you stand for — know what is important to you. Every religion in the world shares one ethical Maxim — the 'Golden Rule'. Whether you are religious or not this rule has enormous practical value. It is 'Do to others as you would have them do to you'.

When we are young we are totally dependent on others. As we mature we gain independence. Some people never progress beyond that state. However, the highest state is a recognition of interdependence and that depends on shared values.

Values are based on ethics and are expressed (partly) through our rules of conduct. However, people do confuse true/false distinctions (logic) with right/wrong distinctions (ethics) and good/bad ones (quality). Recently, on the radio I heard a distinguished Bishop (*Today Programme, Radio 4*) say that wrong is the opposite of true! He was arguing that there are absolutes and he was against relativity. The issue here, for me, is that just because there may be no absolutes does not mean everything is relative — far from it. Many of the truths to which we refer are pieces of information that have a dependability. It is our professional duty to make sure that our decisions are based on sufficiently dependable information. The ultimate test is in the courts which are firmly in world 3.

We have a duty of care to:

- be informed and up to date;
- think about the consequences of what we do;
- be open and communicate openly;
- and these are all duties we can develop through our professional Institution.

Thus, we can leave the business of absolute truths to the philosophers and focus on how we accomplish things and justify what we have done. We build structures in spite of not knowing everything there is to know. We have a duty of care to check out all that is relevant. Participation in Institution discussions is part of demonstrating that duty of care.

We Inject Practical Rigour Using Practical Intelligence

Engineering method is often criticized by pure scientists as non-rigorous because we use approximations and judgement. In fact, structural engineers *have to be rigorous* for two reasons. Firstly, because our structures will be subject to the ultimate test — that of nature. Secondly, because we have to justify our decisions in a society that increasingly questions expertise.

Practical rigour derives from practical intelligence. Thanks to the tremendous advances in brain science we can now observe practical intelligence at work. Research [9] has shown that connections between neurons in the brain develop as someone learns to play the piano. This is a clear proof that practical skills develop into practical intelligence, i.e. brain power.

Rigour is the strict enforcement of rules to an end. Mathematical logic is the ultimate form of absolute rigour: it has one value — truth. Theorems are deduced using axioms (rules) which are true by definition. Physical science aims at precise truth. Here truth is a correspondence to the facts, which, as we saw earlier, has to be set in a context. Practical rigour is much more complex. It is meeting a need by setting clear objectives involving many values (some in conflict) and reaching those objectives in a demonstrably dependable and justifiable way.

As I see it, there are seven elements to practical rigour:

1. *Making it work*: Creating practical solutions to meet explicit needs and delivering a system valued in a variety of ways, not just cost. It includes managing the difference between where we are now and where we want to be in the future. Senge [10] calls this 'creative tension'.
2. *Creating appropriate models*: This involves working with nature — making sensible approximations that respect nature as a cunning adversary. Left alone, the entropy of a system (think of entropy as a level of disorder) will increase and any weaknesses will be found out

and exposed eventually. However, nature is not our enemy, it is our context, our meta-system (Figure 6.5), and so to be sustainable we need to find win–win solutions with nature. Far from being the cause of loss of rigour, as our academic accusers may hold, the approximations of our models are the sources of the practical rigour required to create a solution that meets the needs. Practical rigour requires diligence and duty of care that leaves no stone unturned with no sloppy or slipshod thinking.

3. *Considering the whole as well as the parts*: The scientific approach is one where we look at a problem, break it down into its separate components, take out the difficult bits that we do not know how to solve and focus on what we can solve. It is a process of selective inattention. Practical rigour is the rigour that deals with the bits of the problem that we do not always understand too well. It involves a careful judgement of risk. What is more it involves not just seeing the problem in its separate bits but also how the bits interact to create the whole.
4. *Making judgements*: I have, on more than one occasion, heard engineers refer to their opinions as arbitrary. What can be more calculated to puzzle a member of the general public than that? Professional opinions are not arbitrary, they are based on objective world 3 evidence of varying dependability. Opinion based on experience may be less dependable than measurement or standard theory but it has to be testable against world 3 objective knowledge, ultimately, perhaps, in the courts.
5. *Exercising creative foresight*: Practice requires the creativity to imagine what might happen — how physical things will respond and how people might behave in certain situations. For us it is a deeply intuitive understanding of structural behaviour and risk.
6. *Developing and evaluating dependable evidence*: As I have said, the only clear way to judge the dependability of evidence is to subject it to as many tests as seems appropriate. For hard systems the tests are essentially based on measurement. For soft systems the tests are essentially based on rigorous discussion. Measurement is easy for simple things like length and weight. There are few agreed processes for soft systems (one such is voting [6]). Essentially, the main form of test for totally 'soft' information is to subject it to intense discussion, looking at it from all points of view and examining all implications. This is quite different

from showing information is precise and true — it is a justification that can be quite fuzzy and incomplete yet be enormously useful.

7. *Feedback and learning*: One of the seven habits of successful people identified by Covey [11] is learning to improve or self-renewal. However, as an industry we are not as good at learning from projects and from the way our structures behave in use as we should be. This is a crucial role of the Institution.

Quality Should Include Everything that is Wanted: Phil's Maxim No. 5: 'Take a Whole View of Quality'

So often I have heard engineers refer to quality, safety and economy — but when they say quality I think they usually mean functionality. Safety and economy are part of quality. The phrase should be functionality, safety and economy and much else besides. We need to ask the question for every project with which we get involved: 'What is quality starting right at the top of the supply chain, i.e. with the customers of our clients. How does the asset we are helping to provide relate to the needs of the client's customers and hence to the client's business. How are we adding value?'

Note that because key performance indicators (e.g. cost, predictability of cost, number of accidents, number of defects at handover, staff morale, client satisfaction, etc.) are important enough to be tracked then they express that to which we give worth — they are values. Seen this way, hard system variables such as span, yield strength, and Young's modulus are also values.

Phil's Maxim No. 6: 'Add Value to Others and You Will Add Value to Yourself'

Most of us have been educated to believe in a world of facts and objective truth. We recognize issues such as team spirit, safety culture and shareholder behaviour but we think of them as subjective when they are actually in Popper's world 3, i.e. they are objective. But measurement is problematic and the results may be not as dependable as we would like but they are not arbitrary. They are useful if we see them for what they are. It is absolutely crucial that we learn how to add value using them.

Social capital as referred to earlier is an example. To do this we must think win–win. This is not a technique — it is a philosophy of human interaction. Some people find it hard to square win–win negotiating with continuing to have a competitive advantage. Competitive advantage comes from differentiation in the market place. The essential idea is to focus on adding value to your client and his customers in the value chain. It comes not from keeping things to yourself but realizing your interdependence with others.

Phil's Maxim No. 7: 'Think Systems'

I believe that we have consciously to adopt a systems approach [6]. The systems approach derives from biology. There are three key ideas in my approach to systems:

(a) wholes and parts, with emergent properties;
(b) connectivity;
(c) total process.

Let me explain. The systems thinker sees the world in layers or levels. Starting at the human scale and looking inwards we see down to atoms, sub-atomic particles, even down to super strings. Looking up to the cosmos we have the stars, the constellations and the vastness of space. The traditional 'purist' view sees the top and bottom as 'fundamental' because it stretches our horizons. But are these extremes worth any more to us? The answer is that it depends on what we want (the answers to *why* questions) and that depends upon our values. If we want a bridge then superstring theory is of no value. If we want true precise knowledge we must search ever deeper and higher.

One of the single most important systems ideas is that of a property that emerges from the co-operation of parts, thus making the whole more than the sum of its parts. For example, your ability to walk 'emerges' from the co-operation of all of the subsystems in your body. Process holons are a peg on which all attributes can be hung (Figure 6.6) [6].

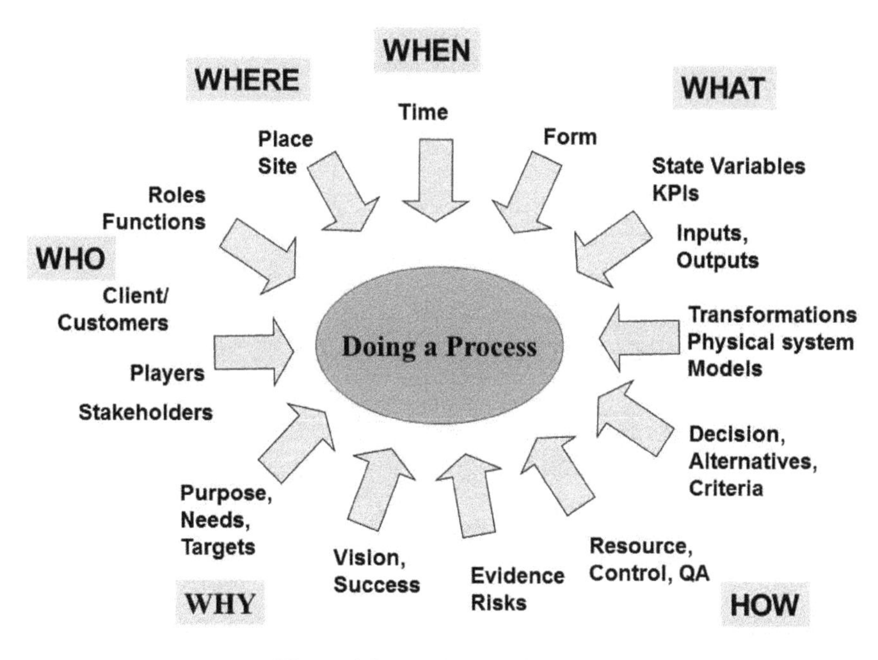

Figure 6.6. Attributes of process.

It is helpful, I think, to see the unity and the diversity of the wholes and parts in our somewhat fragmented industry if we are going to achieve the changes we want to see. Let us hope that the new Strategic Forum chaired by Sir John Egan will fulfil this role.

Figure 6.7 shows the top level Business, Customer, Integrating, Operating, Delivery + Regulatory (BCIOD+R) analysis of our industry following the methods of Blockley, Godfrey [6].

Phil's Maxim No. 8: 'Biology will be the New Engineering Science'

Figures 6.8–6.10 demonstrate the well-known idea that nature has always been an inspiration to designers. The Gare do Oriente in Lisbon, the Ling Shan Brahma Palace Meeting Room in Wuxi, China and the Lotus Temple Delhi, India are good examples.

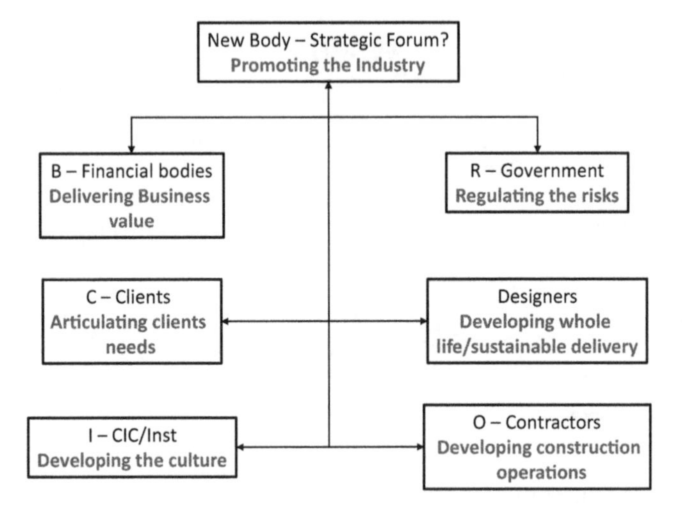

Figure 6.7. Promoting the industry.

Figure 6.8. Gare do Oriente, Lisbon — Calatrava 1998.

Source: Reproduced with the permission of Bobo Boom, CC-BY-SA 2.0 via Wikipedia Commons.

Figure 6.9. Ling Shan Brahma Palace Meeting Room Wuxi China.

Equivalent Systems

We can learn a lot from an analogy between wholes and parts (between the processes and sub-processes) in artificial and natural systems (Figures 6.11 and 6.12). A very effective way to explain to members of the public what structural engineers do is to say to them: if the building team were producing the human body then structural engineers would be designing the muscles and skeleton. The figure also highlights other roles.

Figure 6.10. Lotus Temple Delhi, India.

Source: Reproduced with the permission of Vandelizer, CC_BY_SA 2.0 Wikipedia Commons.

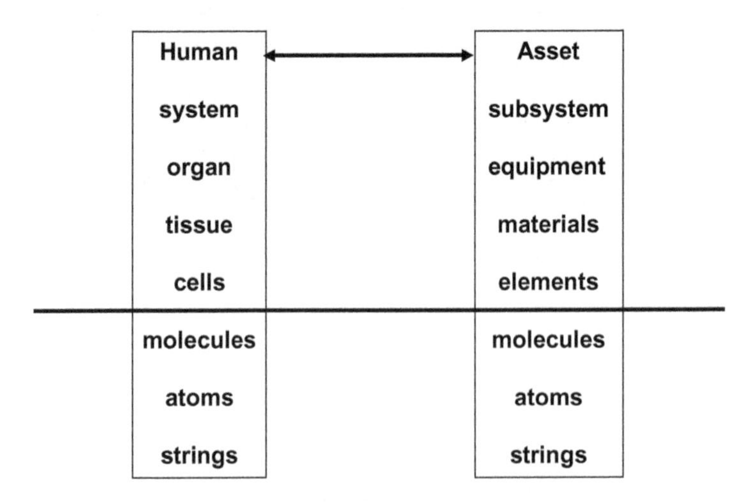

Figure 6.11. Systems levels of definition.

Systems

Perhaps the greatest single lesson from biology and from modern developments in non-linear systems analysis is that it is possible to obtain very complex behaviour from interacting simple ones. This was highlighted in

Body	Body	Building
Systems	Some Organs	Systems
Musculoskeletal	Bone, muscle	Structural
Integumentary	Skin	Cladding
Circulatory	Heart, blood	Circulatory
Respiratory	Nose, lungs	Environmental
Digestive	Stomach, pancreas	Energy
Urinary	Kidneys, bladder	Waste
Immune	White blood cells	Safety, security
Nervous	Brain, nerves	Information
Endocrine	Hormones, glands	Regulation
Reproductive	Tested, ovaries	?

Figure 6.12. Equivalent systems.

the 2000 Maitland Lecture by Professor Susan Greenfield, in which she emphasised the crucial role of connectivity between neuron cells.

Chaotic behaviour [12] is an emergent property which can result in a severe limit on our ability to predict future behaviours. For a two-hinged pendulum with a vibrating fulcrum, two trajectories that start out with only marginally different initial conditions behave totally differently after a few cycles. Biomimetics (literally mimicking biology) is a new subject [13, 14]. The aim is to learn from nature to change the way we do things. Some of the possibilities for engineering are as follows:

- self-designing and repairing structures;
- clever-folded structures, since in nature shape is cheaper than materials;
- structures that adapt to changing forces and circumstances resulting in forms that are tolerant of error and are hence robust (primitive examples of adaptiveness are earthquake structures with active damping devices);
- nanofabrication that will produce materials that can change and adapt [15];
- developing some principles [13] for sustainability since nature:
  - runs on sunlight;
  - uses only the energy it needs;

- fits form to function;
- recycles everything;
- rewards co-operation;
- banks on diversity;
- demands local expertise;
- curbs excesses from within;
- taps the power of limits.

Here are Some IStructE Messages

The Institution has six main constituencies: members, the general public, children, students, government and other professionals. We have many things to say to them, both in overview and detail, but I believe that we should have some simple clear messages that we have all agreed we want to communicate to each of these constituencies. I am now going to make some suggestions — I will look forward to discussing these during my visits to branches and divisions around the world.

To Members, I Say

- get involved in 'Rethinking Construction' — our task group chaired by Alan Gilbertson will tell you how;
- identify and communicate the enormous value you add to your clients — do it very specifically each time you do something — i.e. use Phil's Maxim No. 2 and think about the practical rigour you bring to the team;
- nurture your clients — make sure that your appearance, the way you speak at meetings or on the phone and the way you write reports is the most effective — if necessary take specialist advice on this. Broaden your expertise so you can relate what you do to wider issues;
- help make our web site work for us all — see my article in the journal [16];
- use the biological metaphor to explain to non-specialists what you do.

To the General Public, Together Let Us Say

- structural engineers 'make it stand up — safely';
- If the construction team was building the human body we would be creating the muscles and the skeleton on which all else is hung;
- see how much this person saved because he employed a chartered structural engineer.

To Children, Together Let Us Say

- look at the sheer scale of some of the structures we build and the impact on our surroundings;
- look how engineers are leaders and managers;
- see the lawyers sitting round the table with the structural engineer — they are hanging on every word as the engineer explains some technical point.

To Undergraduate Students, Together Let Us Say

- the industry is changing and there are new opportunities through 'Rethinking construction';
- prepare yourselves with the knowledge and skills to take them.

To Government, Together We Say

- we are the experts, we are highly qualified up-to-date professionals, you need us to protect public safety;
- make us a legal necessity;
- look at all of this research — it shows you need chartered structural engineers involved by law to exercise your duty of care. Mistakes by people doing work for which they are not properly qualified can cost you a lot of money.

To Other Professionals, Together Let Us Say

- see how by paying this engineer more you saved the client a lot of money;

- you need us for our practical rigour (we will have to explain what we mean, of course);
- together we must explain to the media that being wrong is not the same as being negligent.

To the Institution's Executive and Elected Officers, Together Let Us Say

- get involved with the media, invite them to the Institution, have regular briefings;
- provide media training for all senior executives, elected officers including branch chairs and secretaries;
- comment immediately on events — especially controversial ones, put it on the website and send it to the media;
- make us put behind us the image of the 'gentleman's club' and become a dynamic force for change.

Remember that to change others we must change our behaviour towards them. If we want to influence the perceptions that clients have of us then we have not just to add value — we have to demonstrate to them that we add value. If we want to influence public opinion simply explaining what we do is not enough we have to influence the values that both groups hold.

Engineers lack a human face to the public. The public see us often as dealing with prescribed things — applying science with little room for judgement and little room for human values. We need to demonstrate the human face of engineering.

Conclusions

My objectives are reviewed in Table 6.1 with some conclusions. Engineers are naturally concerned over lack of status and public recognition. We must recognize that to change others you must change yourself. I have set out some of the things that I believe we have to do together. I look forward to discussing these ideas in more detail with members over the next year.

Table 6.1. Objectives and conclusions.

Objectives	Review
Articulate the value of the institution	The carrots are: tough exam, learn best practice, be aware of developments, good for you. The stick is that you may be required to show duty of care in court.
Argue for new perspectives and skills	Show we can think out of the box, use systems thinking: highly connected interacting processes which are parts and wholes; understand interdependencies.
Understand practice in theory better	Start with why questions, answers drive all else, understand practical rigour.
Look forward to some new research futures	Biology will be the new engineering science.
Suggest some IStructE messages	Let us propagate simple shared messages together and repeat them whenever possible; e.g. structural engineers make it stand up — safely.
Draw some conclusions	Phil's 8 Maxims include: look outside in and inside out; focus on adding value to client and stakeholders.

Acknowledgements

I want to thank all my friends and colleagues both inside and outside the Institution for the many discussions that have helped me formulate my ideas over the years. In particular, I thank Simon Pitchers for reading the first draft of this address and making several penetrating observations that have significantly improved the text.

References

[1] Pugsley, A. G. See https://en.wikipedia.org/wiki/Alfred_Pugsley (Accessed on June 2019).

[2] Banbury, K. (2000). *Rethinking the Institutions*, Report to Inst Civ. Engrs., London.

[3] Putnam, R. (2000). *Bowling Alone*. Simon & Schuster, New York.

[4] Blockley, D. I. (1983). Phil's 8 Maxims. *Struct. Eng.*, 61A, No. 8, 292–294.

[5] Egan, Sir J. (1998). *Rethinking Construction*. Department of the Environment Transport and the Regions, London.
[6] Blockley, D. I. and Godfrey, P. S. (2000). *Doing it Differently*. Thomas Telford, London.
[7] Popper, K. (1976). *Conjectures and Refutations*. Routledge and Kegan Paul, London.
[8] Magee, B. (1973). *Popper.* Fontana Modern Masters, London.
[9] Pascal-Leone, A., Dang, N., Cohen, L. G., Brasil-Neto, J. P., Cammarota, A. and Hallett, M. (1995). Modulation of muscle responses evoked by transcranial magnetic stimulation during the acquisition of new fine motor skills. *J. Neurophysiol.*, 74, No. 3, 1037–1045.
[10] Senge, P. (1990). *The Fifth Discipline.* Century Business Books, London.
[11] Covey, S. R. (1992). *The Seven Habits of Highly Effective People*. Simon & Schuster, London.
[12] Agarwal, J., Blockley, D. I. and Woodman, N. J. (1998). Safety of non-linear dynamic systems. *J. Struct. Eng.*, 25, No. 1, 37–42.
[13] Benyus, J. M. (1997). *Biomimicry*. Quill, William Morrow, New York.
[14] Vincent, J. (1997). Stealing ideas from nature. Trueman Wood lecture, *Proc. RSA*, London, March.
[15] Drexler, K. E. (1990). *Engines of Creation*. Oxford University Press, Oxford.
[16] Blockley, D. I. (2001). Making the web work for us. *Struct. Eng.*, 79, No. 6, 14–15.

Part III

Understanding Process and Classifying Uncertainty

Preamble

Innovation by thinking out of the box requires us to recognize that no one person has all of the answers — we need to collaborate. There are leaders and followers but it is a myth that only inspired individuals can be creative through a talent so tremendous that some kind of thing (such as a piece of art) springs from their minds fully formed. Creativity is rarely a single act — rather it is often a lengthy process requiring huge amounts of preparation and persistence — even by those who everyone agrees had genius. For example, it seems that Beethoven filled notebook after notebook with musical dead ends and futile variations in painstaking composition. Mozart did not compose his symphonies in one sitting in a single act of inspiration — he considered composition to be an active process, the product of his intellect and carried out under conscious control. In a letter to his father he wrote, 'You know that I immerse myself in music, so to speak — that I think about it all day long — that I like experimenting–studying–reflecting'.

We all need other people to work with and through. Even Mozart and Beethoven needed to collaborate. John Donne wrote, 'No man is an island'. Personally, I like the analogy with sport — successful teams almost always have members that work well together. I really enjoyed playing sport. I played football (soccer) and cricket as a schoolboy regularly with some success including representing Derby Boys at the Baseball Ground — the home of Derby County in those days — an experience that I have never forgotten. Unfortunately, I sustained a severe injury to my left thigh muscle (the *rectus femoris*) in a soccer game for my university hall of residence team that ended my playing career. That was a loss I felt deeply and was not helped by a reporter writing in the student newspaper that 'Blockley was taken off injured after 60 minutes but it made no difference to the game'. I have continued to support Derby County all my life and was able to play cricket and squash well into middle age. The whole of that sporting experience demonstrated to me that a good team of less individually able players can often beat a poor team of more individually able players. It taught me that teamwork and collaboration is about sharing knowledge and skills — it is about 'joining-up' — which in soccer is 'passing the ball'. Teams may have individuals that produce great individual performances but it is the way they join-up that defines great teams. In all areas of life and work individuals may perform well creatively producing bright ideas and moments of inspiration but developing those performances, moments and ideas requires us to work with others. For example, it is well known that creative thinking can come when people bounce ideas off each other through exercises like 'brainstorming' or 'lateral thinking'.

I have said that I see conceiving, designing, building and using infrastructure as processes. I quite purposely use the active headings of the 'ing' form of the present participle in order to emphasize the importance of 'doing' a process. In Part II, we argued that measures of risk are not absolute measures but aids in the process of managing knowledge to control risk. Engineering, we said, is a process of problem solving and decision-making using information with varying degrees of dependability. We outlined a process we called a reflective practice (RP) loop of thinking in action dependent on both practitioner and context.

RP can be formalized through a systems approach based on highly connected interacting and interdependent processes which are parts and wholes. We were introduced to the idea that this kind of approach could be applied to physical systems.

As a consequence, in the 1990s I began to be convinced that everything (including engineering products — inanimate physical systems such as a bridge) has a life cycle — everything exists through time. Consequently, everything is a process. My mistake was not to understand that most people see process as quite distinct from product. They told me I was conflating the two when they are fundamentally different. A product is an object — a 'thing' that is tangible, visible and has a relatively stable form and 'substance'. A process is a series of actions directed towards some end. The idea of a physical object as a fixed unchanging 'thing' is so ingrained in the minds of most people that any suggestion of a product as a process was too much for many to take and they told me that I was simply wrong. As a direct consequence I wrote Chapter 7. With hindsight I should have referred to my idea as 'new process' or invented some completely new name and named existing ideas of process as 'old process' — but I did not. I was just focused on wanting to capture the importance of 'joined-up thinking' because I perceived that so many failures occur through poor joining-up. The difficulty I had in appreciating the opposition to my view of process was that it seems rather obvious that all objects change through time. All I was saying is that change is a process of becoming different — it is just a question of on what time scale. So in geological time rocks have been formed, folded and fractured (i.e. changed) over millennia. In quantum mechanics particles that are more like events than things exist for tiny fractions of a second. In everyday life the man-made objects we use are stable and tangible — like a kettle — but their life cycle of conception, design, manufacture, use and disposal is, to me, self-evidently a process. Natural objects like the sea, a river or mountain range clearly change through time and hence are processes.

People and human systems are complex. Joining-them-up requires us to understand and model the processes that natural systems do organically. These include getting the right information (*what*) to the right places (people *who*) at the right time (*when*) for the right purpose (reason *why*) in the right form (*where*) and in the right way (*how*). I promoted the idea that processes are the basis for ways in which we can integrate the 'joining-up' of disparate systems. I saw a way of formalizing the integrating of hard physical (natural and artificial) systems with soft people and social systems. Although some people could see what I was driving at, very many could not. The strength of belief in the traditional idea that product and (old) process are distinct and different — that (old) process contains the activities that result in a product — prevented many from seeing that those products have a life cycle and thus are themselves (new) processes.

The interacting objects process model (IO) that I introduced in Chapter 5 is a practical manifestation of this kind of systems approach applied to basic mechanics. We decided the best way of testing the idea was to see if we could replicate results from existing methods. Satish Chandra and Jitendra Agarwal were two bright Indian research assistants who came to work with me and my University of Bristol colleague Norman Woodman. Satish eventually went back to India and in 2019 is Head of Department of the Structural Technologies Division of the National Aerospace Laboratories in Bangalore. Jitendra stayed on in the UK and in 2019 is a Senior Lecturer in the University of Bristol. Norman was the finite element expert who contributed some very fundamental ideas. The methodology we developed was completely new. We translated structural dynamic finite element analysis into a systems framework where elements are interacting process holons. The methodology is powerful for two basic reasons. First, we use parallel processing (on a massively parallel connection machine computer) which is efficient. Second, we created much greater flexibility in the way behaviours are characterized. We demonstrated a relatively straightforward way to simulate a physical process. However, our main conclusion was that it worked.

In Chapter 7, I argue that systems thinkers start from change. They realize that everything has life cycle of change — but change that is set in the context of a system containing other processes — some at higher and some at lower levels of definition. My **Learning Point No. 7** is that 'change is a "new" process of becoming different'. Of course, physical processes do not have purpose. The important nuance here is that as we interact with natural and artificial physical processes we imbue them with purpose. We act with a purpose which will vary from 'gaining understanding' to 'making a computer' or 'building a bridge' and consequently see things (model) through the 'spectacles' of that purpose. It is fairly clear that artificial 'things' have a purpose that we give them but the purposes we give to natural systems is trickier. The answers lie in the question why. Why do objects fall to the ground? Our answer is gravity. Why does electrical current flow? Our answer is that potential difference or voltage causes current or amps to flow. But there is a duality in that flow creates potential. So, whilst human purpose arises from human intentionality in solving human needs and wants so physical processes arise from human-derived concepts of potential and flow. So as long as we understand purpose as deriving from answers to questions 'why' then we have a common basis to think about hard physical and soft human systems with one theory.

With all that in mind then **Learning Point No. 8** follows. It is that 'Purposeful new process is the key to integrating people, purpose and "old" process'.

In Chapter 6, I introduced the idea that *why* = *how* (*who, what, where, when*). In other words, questions *why* capture the potential, the purpose, functionality, success and failure criteria, concerning a given process at a given level of definition. Answers to all *how* questions capture the transformations or methods that will change the starting or input state to a finishing or end state in a defined time period. Answers to all of the *who, what, where and when* questions determine the flow created by the potential. Within the flow questions in particular the *who* questions define changes in the players or actors, clients and stakeholders. Answers to *what* questions define changes to all of the state variables and performance indicators. Answers to all *where* questions define place and context. Answers to all questions *when* define parameters of time. These may be fast (as in sub-atomic physics or the flutter of a bridge deck) or slow (as in climate change or geological eons). The formulaic expression is not to be interpreted as a piece of mathematical algebra, rather it is an attempt to represent the idea that *why* is the voltage or difference of potential that drives the transformation *how* in the flow or current of change in the attributes *who, what, where* and *when*.

The three ideas at the heart of delivering systems thinking are thinking in layers, thinking in connected loops and thinking about *new* processes. These are set in the context of a system containing other connected processes — some at higher and some at lower levels of definition. So, **Learning Point No. 9** in Chapter 7 is that 'systems thinkers build models based on three basic concepts: (1) layers of, (2) connected, (3) new processes'. All processes have attributes that are characterized using *why, how, who, what, where, when*.

But more than that as we saw in Chapters 5 and 6 some properties of process holons 'emerge' from the interactions. Emergent properties are those that result from interactions between the parts that make the whole. The whole is more than the sum of its parts. **Learning Point No. 10** is that 'interactions between processes create emergent properties or characteristics'.

Systems thinking is not simply an engineering approach, rather it is a philosophy for solving many practical problems such as joined-up government, social work, dealing with climate change and terrorism.

But what about uncertainty? As I said at the start of this book there is uncertainty in everything we do as human beings. In Chapter 8, I identify six kinds of uncertainty under the headings of truth, trust, clarity, changeableness, incompleteness and risk. I argue that truth and changeableness (together as one kind) with clarity and completeness are three central attributes of information and that trust is central to the making of decisions by 'experts'. They all are ingredients of the main task of managing and controlling risk. Reductionism has been successful for the analysis of hard systems but soft systems are governed by the behaviour of people which is so complex as to be hard to define and difficult to analyze. The emphasis in soft systems is not on prediction but rather on managing a process to achieve desired outcomes based on dependable evidence which is grounded in a context. **Learning Point No. 11** is that 'soft systems are difficult to predict — use dependable evidence to achieve purpose'. In recent years, risk studies have come to the fore. Two assumptions, commonly made in theoretical risk and reliability analyses, have a long history. The first is that uncertainty is either *aleatoric* or *epistemic*. The second is that standard probability theory is sufficient to express uncertainty. **Learning Point No. 12** in Chapter 8 is that 'the three orthogonal structural attributes of uncertainty are fuzziness, incompleteness and randomness'. Chapter 8 also incorporates an extract from a paper I wrote in 2013 containing a conceptual analysis of uncertainty. I conclude that categorizing uncertainty as either *aleatoric* or *epistemic* is unsatisfactory for practical decision-making. I argue that uncertainty emerges from three conceptually distinctive and orthogonal *structural* attributes FIR, i.e. fuzziness, incompleteness (epistemic) and randomness (aleatory). Aspects of uncertainty, such as ambiguity, dubiety and conflict, are complex *interpretations* of the structural mixes of interactions in an FIR space. By structural I mean 'being constructed of — or the arrangement of parts of' in order to distinguish it from words that interpret meaning. To manage future risks in complex systems it will be important to recognize the extent to which we 'do not know' about possible unintended and unwanted consequences or unknown unknowns. In this way, we may be more alert to unexpected hazards.

Chapter 7

The Importance of Being Process*

Abstract

The purpose of this chapter is to outline the particular interpretation of systems thinking developed at the University of Bristol over the last 30 years. The importance of process and uncertainty are central themes. Put at its simplest, systems thinking is joined-up thinking. It is getting the right information (*what*) to the right people (*who*) at the right time (*when*) for the right purpose (*why*) in the right form (*where*) and in the right way (*how*). The three ideas at the heart of delivering systems thinking are thinking in layers, thinking in connected loops and thinking about *new* processes. Everything has life cycle and hence is a process — but one that is set in the context of a system containing other connected processes — some at higher and some at lower levels of definition. All processes have attributes that are characterized using *why*, *how*, *who*, *what*, *where*, *when*. There is a need to integrate hard and soft systems. This requires us to be very clear about the meaning and usage of the terms subjective and objective when we argue that engineering judgement is both valid and important. It is argued that truth is to knowledge as the inverse of risk is to action. The three *structural* attributes of uncertainty are stated as FIR — fuzziness, incompleteness and randomness. Robustness and its inverse, vulnerability, are crucial,

*This chapter was originally published in *Civ. Eng. Environ. Sys.*, 2010, 27, No. 3, 189–199.

though often ignored. Systems thinking is not simply an engineering approach, rather it is a philosophy for solving many practical problems such as joined-up government, social work, dealing with climate change and terrorism. Finally, it is argued that our journey to 2030 requires us to adopt an evolutionary observational approach using systems thinking.

Introduction

In the play 'The Importance of Being Earnest' by Oscar Wilde premiered in 1895, Lady Bracknell has an exchange with the leading character, Jack Worthing, who wants to marry her daughter. After Jack tells her he was found abandoned as a baby in a handbag on Worthing station, she exclaims, 'A handbag!' The line has been delivered with such a mixture of horror, incredulity and condescension by so many famous actresses, that it has become a classic line in English-speaking drama. Lady Bracknell continued, 'To lose one parent, Mr Worthing, may be regarded as a misfortune; to lose both looks like carelessness'.

To most people a handbag is an object. To a systems thinker a handbag is a process that plays a role in other processes. In simple terms, the handbag has a life cycle in which it was conceived, designed, made and used and eventually disposed of. However, it is unlikely that it was designed to carry an abandoned baby. Nevertheless, in that role or function the handbag played a crucial part in keeping the infant Jack safe and, as revealed as the play unfolds, allowing him to be found and rescued into a good family upbringing. So, another idea at the heart of systems thinking is dealing with uncertainty — managing the unexpected — in particular recognizing that there are limits to what we know and hence what we can predict. Richard Whatley, a 19th century Irish cleric said, 'He who is not aware of his ignorance will only be misled by his knowledge'. In a similar vein from an unknown source, 'Once you have accumulated sufficient knowledge to get by, you are too old to remember it!'

The importance of process and uncertainty are central themes of this chapter. The objectives are (a) to review in outline the particular interpretation of systems thinking developed at the University of Bristol over the last 30 years, (b) to sketch out some examples and (c) to share some thoughts about future directions.

What is Systems Thinking?

So how do we characterize systems thinking as developed at Bristol (for simplicity I will drop the phrase as developed at Bristol in the rest of the chapter and assume it is implied). Put at its simplest, systems thinking is joined-up thinking. By that I mean 'getting the right information (*what*) to the right people (*who*) at the right time (*when*) for the right purpose (*why*) in the right form (*where*) and in the right way (*how*)'. When systems lack joining-up then a message does not get sent or received or is poorly formulated, incomplete, misleading or is without adequate justification.

There are three ideas at the heart of delivering systems thinking. They are thinking in layers, thinking in loops, thinking about *new* processes. Let us consider these in turn.

First, a systems thinker sees the world in levels of definition — all of which are important and useful depending on the type of problem being addressed. None are deeper or more fundamental than any other. You choose a level based on your need to solve a problem. High-level statements tend to be less precise but of wide scope — a national authority might assert that 'All bridges in the UK are safe'. At a lower level, a bridge designer might say that 'The stresses in all arch bridges are low'. At a detailed level an engineer could calculate that 'The deflection at the centre of the span of this bridge was a maximum of 50 mm under live load'. Lower-level definitions tend to be of narrower scope, be more precisely stated and in reductionist science be considered as more fundamental.

The word 'holon' was first suggested by Koestler [16] to capture the idea that at a given level of definition something is both a whole and a part. That something is a 'whole' in the sense of it being a totality, an individual thing made of parts at a lower level of definition, but also a 'part' in the sense of playing a role (just like the 'handbag' referred to by Lady Bracknell) in a higher-level set of processes.

Second, a systems thinker is a 'loopy thinker'. Systems thinkers look for connections and feedback and feed forward. Many people tend to think in straight lines — moving from cause to effect. Senge in his excellent book [22] on systems thinking gives many examples of thinking in loops.

Connectivity or connecting (i.e. joining, linking and communicating) is at the heart of modern complexity theory, leading as it does to the important concept of 'emergence'. An emergent property at a level of definition is one that results from interactions between the parts that make the whole. It is in this sense that the whole is more than the sum of its parts. So, any holon has emergent properties. For example, the air in a balloon has pressure. If we examine this 'whole' balloon in terms of its parts at a lower level of definition then one contributory part is the air pressure. At this lower level, the air can be modelled as molecules of oxygen, nitrogen, etc. 'buzzing around' and colliding with each other and with the inside surface of the balloon. Such a model, based on statistical mechanics, depends on capturing the interactions between the constituent parts which make up the pressure as defined at the higher level. Another example of emergence in construction management is in the use of critical path analysis. Here, a network represents the time interdependencies between activities that are needed to bring a project to a successful conclusion. Those time attributes such as earliest and latest start times emerge from the interactions between activities. So from this we see that 'interactions between processes create emergent properties or characteristics'.

Third, a systems thinker sees everything as a 'new' process. The adjective new is included here to emphasize the need to reject all existing preconceptions of what constitutes a process and to create a new and all-encompassing definition. So, a new process is not just an input being transformed to an output, or a Gantt bar chart, a recipe, a flowchart, a network of an IDEF0 diagram — it is all of these and more. A new process characterizes everything that we know. As already stated, everything is viewed and modelled as a process because everything exists through time. The handbag to which Lady Bracknell was referring, a kettle, a building, an aeroplane, a power station, an airport terminal and all living things, including human beings, are all represented as processes. So, a systems thinker starts by knowing that everything has life cycle — but one that is set in the context of a system containing other processes — some at higher and some at lower levels of definition. All processes have attributes that are characterized using *why, how, who, what, where, when* (Figure 7.1).

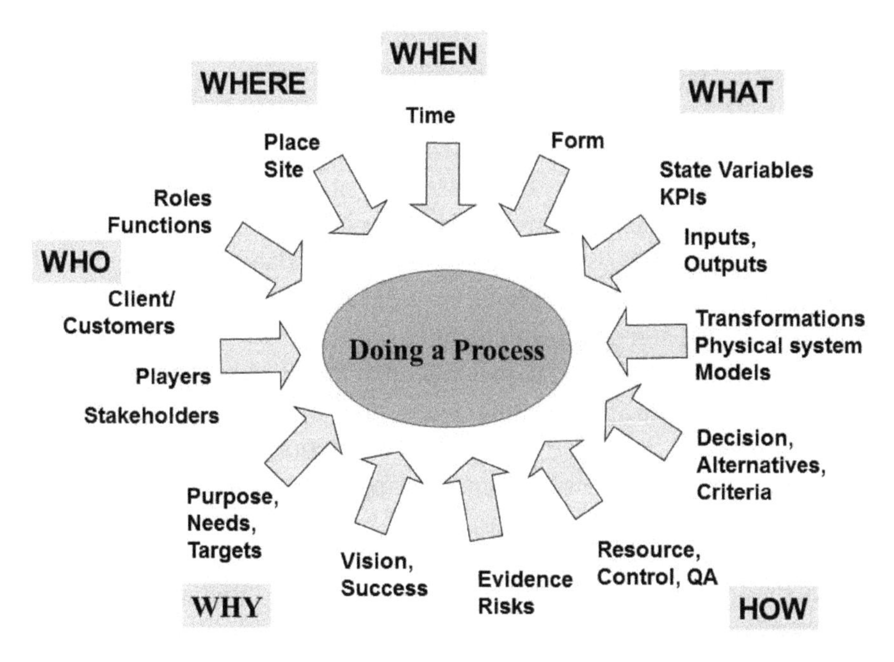

Figure 7.1. Thinking about NEW process.

The purpose, functionality, success and failure criteria, concerning a given process at a given level of definition, derive from all the relevant questions *why* that can be identified. Answers to all *how* questions define the transformations or methods that will change the starting or input state to a finishing or end state in a defined time period. Answers to all *who* questions define players or actors, clients and stakeholders. Answers to *what* questions define all state variables and performance indicators. Answers to all *where* questions define place and context. Answers to all questions *when* define parameters of time. These may be fast (as in sub-atomic physics or the flutter of a bridge deck) or slow (as in climate change or geological eons). We can capture the relationships between these questions as *why* = *how* (*who, what, where, when*). However, this expression is not to be interpreted as an algebraic formula — rather it is an attempt to represent the idea that *why* is the voltage or difference of potential that drives the transformation *how* in the flow or current of change in the attributes *who, what, where* and *when*.

There is one crucial state variable or *what* attribute that needs to be highlighted. This is a measure of the dependability of the evidence that the process will be successful. As stated earlier, the meaning of success is formulated from the questions *why*. This success is normally expressed in terms of target states, aims, objectives and, at higher levels, by mission and vision statements. In work at Bristol, we have used Italian flags based on interval probability as this measure of the dependability of the evidence that the process will be successful or will fail [6]. The measure is used both in hard and soft systems thinking.

Hard systems are physical, material set of things — but here conceived and modelled as processes. A soft system involves human beings. In hard systems, there is an action that creates a reaction which we normally understand through reductionist engineering science. That understanding is sufficient to create hard systems, such as bridges, that work or are fit for purpose, i.e. they are successful. However, all hard systems are embedded in soft systems [8, 11] since it is through soft systems that we understand the world around us. But soft systems are hard systems with an extra complication — multiple layers of human intentionality. At its simplest, intentionality is having a purpose, aim or goal. It is this multiple-layered interacting intentionality that makes soft systems so difficult. Table 7.1 shows the potential and flow and impedance in some typical hard and soft systems.

Table 7.1. Thinking about new process — Potential and flow.

System	Potential	Flow	Impedance
Electricity	Volts	Amps	Resistance, capacitance, inductance
Mechanics	Velocity	Force	Damping, mass, flexibility
Water pipes	Pressure head	Flow	Drag, open tanks/reservoirs, closed tank?
Traffic	Need	Flow	On-street parking, off-street parking, route changes
'Soft'	Why — Creative tension	(Who, What, Where, When)	Ambiguity/conflict, capacity to perform, capacity to adapt/innovate

Subjectivity and Objectivity

A systems thinker must be a philosopher by night and a man of action by day — a reflective practitioner [3–5, 14, 21]. By that I mean that systems thinkers think that thinking about thinking is important and that it is important to develop as clear a set of ideas as is possible to underpin what is accomplished by practical action. In this regard, the works of Popper [20], Kuhn [17] and Dewey [13] are influential. The first realization that this reflective thinking reveals is that we must recognize that science cannot deliver absolute Truth. Truth with a capital T is an attribute of statements that correspond to facts in all possible contexts. Indeed, as Popper pointed out, what we mean by describing something as a fact is itself problematic and has to be defined in a meta-language. This leads to a (philosophically) uncomfortable infinite regress. Systems thinkers resolve this by seeing all statements as only being true in a specific (but often not clearly identified) context. They therefore use a notion of *dependability* to mean common sense contingent truth. A dependable statement is one that is highly tested — where the tests are to some degree sufficient but not necessary since non-tested statements may sometimes be dependable. Tests will vary from almost nothing (and rely instead on an assessment of the trustworthiness of the messenger) to very strong where we depend on well-established engineering science. Nevertheless, all practitioners know that, again following Popper, the potential for unintended and unwanted consequences are ever present. So, systems thinkers do not focus purely on the predictive tools of reductionist engineering science, rather they use the tools of science in the context of evidence towards the management of a process to success. Evidence comes from past experience (such as historical data and case studies), present observations (such as the growth of a crack or deflection of a retaining wall) and future possibilities (such as predicted by finite element analyses). Any piece of evidence from whatever source is used as appropriate. The sources are therefore many and various from 'subjective' judgement to 'objective' measurement. Each piece of evidence has to be understood in context and its dependability assessed. Inevitably this requires a comparison of what seems to be 'chalk and cheese' and calls for a very clear idea of what constitutes subjective and objective evidence. Blockley and Godfrey [8] have discussed the use of hard and soft measures in some detail following an earlier analysis by the author [2].

Subjectivity is a difficult concept and widely disparaged in the 'hard' sciences. The consequence is that the skill of judgement tends to be undervalued by academics and researchers. Systems thinkers place great emphasis on being clear about the meaning and usage of the terms subjective and objective — the thinking is as follows. We reach out to the world through our senses. We hear noises, we see things and we touch them. But what is the world really like? If our eyes were sensitive to X-rays we would see the world quite differently. Subjective means 'existing in the mind or belonging to the thinking subject rather than to the object of thought'. So, our mental models would be quite different if our senses were different. Subjective perceptions are of two kinds. First, we have perceptions and private thoughts that cannot be shared — they are truly subjective, e.g. a pain in my stomach. Second, we have many perceptions that we can share. Indeed, by discussion, we agree about them. We call these shared perceptions inter-subjective. An example might be the colour of a fabric. We have no way of knowing the actual perceptions of others but we can agree that each time we perceive that colour we all use the same name. In that way we learn to describe the inter-subjective perceptions that we all have. With these shared perceptions we construct ideas and relationships. We make measurements in ways that are repeatable and dependable. As we agree, we begin to describe the knowledge as objective and when it is also testable we call it science. Objective information therefore exists outside any one individual person's mind [20].

Unfortunately, objectivity is often wrongly thought to consist of only measurable information. Of course, measurable information is objective but not *vice versa*. Objective is commonly defined to mean belonging to the object of thought rather than to the thinking subject. However, we can only construct such objects through shared inter-subjective perceptions that clearly exist outside of the mind of any one individual — it is information that has been agreed and is available to everyone. We can think of it as all of the books in the library and all of the information available on the Internet. Objective knowledge has an objective existence (i.e. outside any one mind) even though it derives from our collective subjective minds. However, this test of shared existence of knowledge must not be confused with the test of whether it is true or false. Objective knowledge can be fictional, e.g. Sherlock Holmes and mermaids exist objectively but are fictional. Objective information can be true or false, accurate or inaccurate, dependable or undependable.

Systems thinkers value engineering judgement because it relies on experience and personal characteristics and while it is not easily measured and demonstrably dependable, except in hindsight, it is essential for good decision-making. It is not arbitrary and subjective, as so often is asserted, though when it is measured (for example, by voting) the results may be of variable dependability. But how do we recognize good judgement? Dependable judgement has to be tested if at all possible and there are various ways that can be done. The least satisfactory way is to rely on the previous sound performance of the decision maker. The best way is to test the judgement against specific criteria or physical circumstance — but this is often not possible.

Clearly systems thinkers, indeed all decision makers, prefer dependable objective knowledge but they do not reject engineering judgement based on experience and expertise. However, they do tread carefully as they attempt to manage the process, about which decisions are being made, towards success and avoiding failure.

Knowledge and Action

Systems thinkers see developing knowledge and action as leapfrogging over each other. In other words, to act you need to know and to know you need to act. Reductionist philosophy sees knowledge as more fundamental than action but systems thinkers value both equally. Indeed, truth is to knowledge as the inverse of risk is to action as illustrated in Table 7.2.

Table 7.2. Truth is to knowledge as the inverse of risk is to action.

Knowledge	Action
Intention of *knowledge* is to achieve *understanding*	Intention of *action* is to achieve *outcome*
Truth/dependability is attribute of correspondence of understanding with 'facts'	Risk is attribute of lack of correspondence of outcome with *consequences*
Degree of truth/dependability between *true and false*	Degree of risk between *failure and success*

Quality

Systems thinkers see quality as the ultimate expression of what they want to achieve. Quality expresses the totality of what we want from a process. It has two common interpretations, degree of excellence and fitness for purpose which are often confused. Systems thinkers see these two expressions as the same as long as you are clear about context and purpose. Systems thinkers define excellence as a state of pre-eminence or of having the highest value. Value and values are the worth we give to something. This worth is expressed through our purpose — through the answers to *why* and *what* questions. A common example of confusion is demonstrated by the question — 'Which car has the higher quality — a Mini or a Rolls Royce?' A Mini which meets the specification of a Mini perfectly is of high quality in the sense it is fit for the purpose of being a Mini. Likewise, a Rolls Royce which meets the specification of a Rolls Royce is a high-quality Rolls Royce. Thus, in the sense of being fit for purpose, then they are equivalent. However, if we include in our value systems a preference about the degree of excellence of the specification then most of us would agree that the specification of a Rolls Royce is higher than for a Mini. The confusion between quality as excellence and quality as fitness for purpose is often caused by not being specific enough about the values being used. There has to be a clear statement that a Rolls Royce is valued higher than a Mini *if that is what is intended.*

Systems thinking is about integrating all aspects, all points of view and all interests in a given system of multiple interdependent processes to deliver quality — it is therefore important that all important values for all players in a set of processes are clearly identified as early in the overall process as is possible.

Risk and Reliability

Structural reliability theory is a good example of a theory that has been well developed by a number of excellent researchers. However, it is partial and misses much of what is required to improve the management of structural risk. It is based on three misconceptions — firstly, that the only kind of uncertainty is randomness. Secondly, that human error is somehow beyond what needs to be addressed. Thirdly is a neglect, until relatively recently, of robustness and its converse vulnerability — where small damage can lead to disproportionate consequences.

Systems thinkers see uncertainty as having three attributes from which all other attributes of uncertainty emerge. They are FIR — fuzziness, incompleteness and randomness. Some writers refer to randomness as aleatory uncertainty, i.e. dependent on chance, accidental events or other contingencies. At the same time, they refer to epistemic uncertainty as the lack of dependability or truth likeness of conditions under which we can claim to know anything. So aleatory uncertainty is either another word for randomness or we have to unpack exactly what defines an accidental or contingent event since an understanding of those terms is crucial for managing uncertain processes to success. Categorizing uncertainty as epistemic is not very informative. Indeed it is about as useful as stating that all error is human error. The categories are too wide to be helpful. All that we understand is epistemology. The author even asserts that ontology (the study of the nature of existence or being) is epistemology since we can only have an understanding of what it means to exist through how we make sense of the world in which we live. That is why the title of the chapter is the importance of *being* process, i.e. *being* is also itself a process.

Systems thinkers think that attributes of uncertainty such as ambiguity, ambivalence, indeterminacy, unpredictability are emergent properties of FIR. So when we make a statement such as the stresses in arch bridges are low — we say something useful and dependable but not very precise since what we mean by low stress is not precise. We model fuzziness within the levels of definition. At high levels we tend to use fuzzy expressions — such as all the UK bridges are safe. At lower levels we are more precise but we must necessarily reduce the scope of the context in which that precision is applicable.

Incompleteness is perhaps the aspect of uncertainty most neglected — some even deny its very existence. Incompleteness is that which we do not know. Of course, we must distinguish between what we individually do not know and what we collectively do not know. All of us, as individuals, do not know things that others do know, that is why we must work in teams. However, as Plato wrote, 'How can we know what we do not know?'

That is a fundamental challenge to the management of uncertainty and it is one that should make us always work with a degree of humility. There are historical examples of failure due to things which no one knew at the time [6]. To spot them is a real challenge. There seems to be only one Maxim — be prepared. In other words, as we manage processes we develop a habit of looking out for unintended and unwanted consequences and we deal with them before there is serious damage [23]. Unintended but not necessarily unwanted consequence can present new opportunities. Many new discoveries and many new products have been found this way.

Randomness is defined, following Popper, as the lack of a specific pattern in some data. Probability theory handles patterns which occur over populations of data — it is not specific. The interpretation of chance and risk is tricky — we all know of the 90-year-old man who has smoked 40 cigarettes a day since the age of 12. Smoking and ill health are not causally related in the sense that death *always* follows smoking — rather smoking dramatically increases the risk of ill health over populations.

Ambiguity is doubtfulness or uncertainty of meaning or intention — it derives from an unclear, indefinite or equivocal word or expression. It emerges as a potential for more than one interpretation of the meaning of a statement through interacting fuzziness and incompleteness. Likewise, ambivalence is the inability to make a choice or by a simultaneous desire to say or do two opposite or conflicting things again through interacting fuzziness and incompleteness. Contradiction and conflict emerge from incompleteness because they derive from inconsistencies which would not be present in complete information.

Reliability theory addresses only one aspect of uncertainty — randomness. The parameters to a scientific model of a phenomenon are modelled as random variables. Any probabilities of failure resulting from these calculations are more rigorous, but more complicated, versions of traditional safety factors.

It is difficult to include fuzziness and incompleteness in models of the physical phenomena. Capturing these types of uncertainty in the physical models requires us to develop underpinning models at lower levels — rather as statistical mechanics supports thermodynamics. But such models are rarely available — and often, if they are, they are too complex to be practical. Systems thinkers therefore see the best way of dealing with all aspects of uncertainty as follows. Use the best science (explanatory and predictive) available, assess all sources of evidence and use that information to manage the processes in which the phenomena occur in reality to eventual success.

A similar conclusion can be drawn, but even more forcefully concerning human error. Indeed, systems thinkers now reject the term 'human error'. The whole topic is much more subtle, involving as it does all aspects of social science. The capacity for people to do unexpected things seems to be without limit. So again the only pragmatic solution is to gather evidence, systematically and rigorously, about all aspects of a process; and then use that evidence to make good decisions that steer a process through a mine field of hazards to avoid failure limits and to reach success. For example, incubating conditions are hazards, as set out by Turner and Pidgeon [23] and discussed by Blockley [3], that have to be identified and managed.

Vulnerability and Robustness

A much-neglected property of systems is that of robustness and its inverse which is vulnerability. A system is vulnerable and hence not robust when small damage can cause disproportionate consequences. For example, the Ronan Point high rise block of flats or apartments in London was a vulnerable structure, because a small domestic gas explosion in one apartment caused the whole side of the building to collapse in 1968. Vulnerability is about low chance — high consequence risks such as the unexpected collapse of the WTC at 9/11. Good theoretical treatments of vulnerability have been lacking and the topic needs much more attention. This is particularly so for all highly interconnected and interdependent systems from computer networks to engineering structures. A theory of vulnerability, which is also a theory of form, has been developed Lu *et al.* [19], England *et al.* [15] at Bristol. The major purpose is to find weak spots. A system is clustered into a hierarchy of levels of definition. A search algorithm finds various scenarios of possible damage and its consequences. Each is tested against a measure of vulnerability to isolate those which require detailed attention. The importance of this theory is not that such a scenario is highly likely (as would be identified in traditional reliability theory) but that the consequences of small damage are so serious that even a remote possibility must be dealt with.

Solving Practical Problems

Systems thinking is not simply an engineering approach; it is a philosophy for solving many practical problems. Tony Blair's first 'New Labour' UK government wanted to get departments and agencies to work more closely together. They saw a need for better collaboration across organizational boundaries to deal with shared issues. They wanted to improve the flow of information to deliver better services with a focus on the needs and convenience of the customer rather than the provider. So the government introduced major programmes to attempt to modernize many parts of government bureaucracy including the NHS and the criminal justice system. The commitment was set out in the 'Modernising Government' White Paper [10]. One of the three stated aims of the new policy was to ensure 'that policy making is more joined up and strategic'. It was an ambitious project that set out to reform the very processes by which Government itself works. So what happened? The 'joined-up' phrase was used for some time but then got quietly dropped. Perhaps that is understandable since the issues are tough — but they have not gone away. Examples from social work and the criminal justice system abound. There is insufficient space here to give details. However, the terrible treatment and murder of Victoria Climbié was one example that shocked the nation (Department of Health and The Home Office [12]). Harold Shipman was a family doctor who murdered many of his patients (UK Parliament [24]). Ian Huntley was guilty of the Soham murders (Bichard Inquiry Report [1]). All of these case histories are stories where a lack in the joining-up of agencies led to disastrous consequences. Pieces of evidence, considered in isolation, were pieces of a jigsaw. Had the pieces been put together, then a very different picture would have emerged. If someone had been able to do that, then it is quite likely that the tragedies would have been prevented.

Systems thinking is needed for many pressing modern problems such as: managing the consequences of climate change, dealing with terrorism

Table 7.3. Some pressing issues.

Constants that will find new expression	Developing trends	Issues
• War	• Extreme weather	• Ethics
• Crime	• Energy	• Reconciliation
• Gossip	• Waste	• Prediction/control
• Sex	• Technology	• Vulnerability and Robustness
• Misunderstandings	• Poverty	• Work–life balance
• Accidents	• Faith and Reason	

and managing social disorder. Table 7.3 illustrates some other issues including the constants of human history that are always finding new forms of expression such as war, crime and gossip. Developing trends are extreme weather due to climate change; management of energy resources, both fossil and renewable; the management of waste; reducing poverty; and most deeply of all, the apparent conflict between faith and reason. The latter depends on a new radical approach to ethics. Therefore, ethics leads the list of issues [9] together with reconciliation after deep conflict, the use of prediction in control, vulnerability and robustness and work–life balance. Systems thinkers feel that none of these issues will be dealt with satisfactorily without systems thinking.

Finally, Figure 7.2 illustrates a systems thinker's imaginary journey to 2030. It is based on the premise, already stated, that we must steer a path through a minefield of future hazards by being as prepared as we possibly can be. The journey will require us to integrate and manage a set of hard and soft processes as described earlier. The decision-making loop is shown in the diagram — but it is a loop through time to 2030.

So actually it is a spiral in time as we constantly go through the loop to end up where we started but further on in time and a new need to plan. It is an evolutionary observational approach [6, 18].

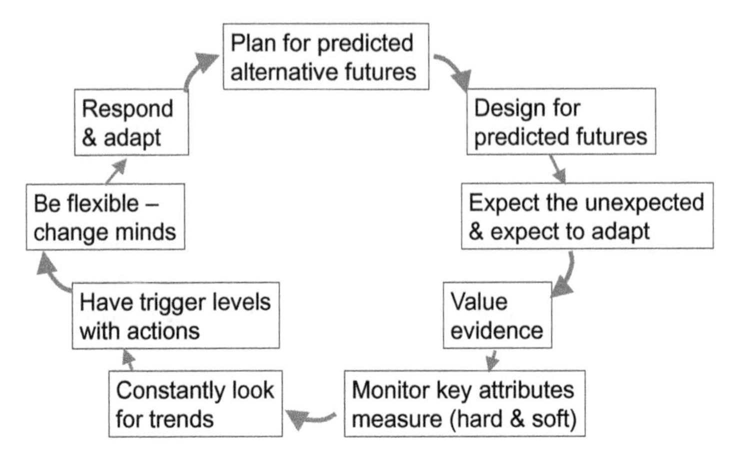

Figure 7.2. Evolutionary observational method.

Conclusions

(1) Put at its simplest, systems thinking is joined-up thinking. It is getting the right information (*what*) to the right people (*who*) at the right time (*when*) for the right purpose (*why*) in the right form (*where*) and in the right way (*how*). When systems lack joining-up then a message does not get sent or received or is poorly formulated, incomplete, misleading or is without adequate justification.

(2) There are three ideas at the heart of delivering systems thinking. They are thinking in layers, thinking about connections and loops, thinking about new processes.

(3) A 'holon', as suggested by Koestler, captures the idea that at a given level of definition something is both a whole and a part. That something is a 'whole' in the sense of it being a totality, an individual thing made of parts at a lower level of definition, but also a 'part' in the sense of playing a role in a higher-level set of processes.

(4) Everything has life cycle and hence is a process — but one that is set in the context of a system containing other processes — some at higher and some at lower levels of definition. All processes have

attributes that are characterized using *why, how, who, what, where* and *when.*

(5) Hard systems are physical, material set of things — but modelled as processes. A soft system involves human beings and therefore has the added complication of multiple layers of human intentionality.

(6) Systems thinkers think that thinking about thinking is important — they are reflective practitioners.

(7) Systems thinkers place great emphasis on being clear about the meaning and usage of the terms subjective and objective. Shared perceptions are inter-subjective. We have no way of knowing the actual perceptions of others but we can agree about them. We use them to construct ideas and relationships. We make measurements in ways that are repeatable and dependable. As we agree we begin to describe the knowledge as objective and when it is also testable we call it science. Objective information exists outside any one person's mind, e.g. all the books in the library.

(8) Systems thinkers value engineering judgement.

(9) Truth is to knowledge as the inverse of risk is to action.

(10) Uncertainty has three structural attributes from which all other attributes of uncertainty emerge. They are FIR — fuzziness, incompleteness and randomness.

(11) Quality expresses the totality of what we want from a process. It has two common interpretations, degree of excellence and fitness for purpose, which are commonly confused but which are entirely compatible.

(12) A much-neglected property of systems is that of robustness and its inverse which is vulnerability. A system is vulnerable and hence not robust when small damage can cause disproportionate consequences.

(13) Systems thinking is not simply an engineering approach; it is a philosophy for solving many practical problems. Tony Blair's first 'New Labour' UK government wanted joined-up government but they failed to deliver. Other examples derive from social work, the criminal justice system, managing the consequences of climate change; dealing with terrorism and managing social disorder.

(14) Our journey to 2030 requires us to steer a path through a minefield of future hazards. It requires an evolutionary observational approach using systems thinking.

References

[1] *Bichard Inquiry Report* (2004). *HC653*. The Stationery Office, London.

[2] Blockley, D. I. (1980). *The Nature of Structural Design and Safety*. Ellis Horwood, Chichester.

[3] Blockley, D. I. (1992a). *Engineering Safety*. McGraw Hill, London.

[4] Blockley, D. I. (1992b). Engineering from reflective practice. *Res. Eng. Des.*, 4, No. 1, 13–22.

[5] Blockley, D. I. (1999). Process modelling from reflective practice for engineering quality. *Civ. Eng. Environ. Sys.*, 16, No. 4, 287–313.

[6] Blockley, D. I. (2010). *Bridges*. Oxford University Press, Oxford.

[7] Blockley, D. I. and Godfrey, P. S. (2000). *Doing it Differently*. Thomas Telford, London.

[8] Blockley, D. I. and Godfrey, P. S. (2005). Measuring judgements to improve performance. *Proc. Inst. Civ. Eng. Civ. Eng.*, 158, 124–129.

[9] Blockley, D. I. and Dias, P. (2010). Managing conflict through ethics. *Civ. Eng. Environ. Sys.*, 27, No. 3, 255–262.

[10] Cabinet Office (1999). *Modernising Government*. White Paper, The Stationery Office, London.

[11] Checkland, P. (1981). *Systems Thinking, Systems Practice*. John Wiley, Chichester.

[12] Department of Health and the Home Office (2003). *The Victoria Climbié Inquiry*. Report of an Inquiry by Lord Laming, Cm 5730, January.

[13] Dewey, J. (1930). *The Quest for Certainty*. George Allen & Unwin, London.

[14] Dias, W. P. S. and Blockley, D. I. (1995). Reflective practice in engineering design. *Proc. Inst. Civ. Eng. Civ. Eng.*, 108, 160–168.

[15] England, J., Blockley, D. I. and Agarwal, J. (2008). The vulnerability of structures to unforeseen events. *Comput. Struct.*, 86, No. 10, 1042–1051.

[16] Koestler, A. (1967). *The Ghost in the Machine*. Picador, London.

[17] Kuhn, T. S. (1962). *The Structure of Scientific Revolutions*. University of Chicago Press, Chicago.

[18] LeMasurier, J., Blockley, D. I. and Muir Wood, D. (2006). An observational model for managing risk. *Civ. Eng.*, 159, No. 6, 35–40.

[19] Lu, Z. *et al.* (1999). A theory of structural vulnerability. *Struct. Eng.*, 77, No. 18, 17–24.

[20] Magee, B. (1973). *Popper*. Fontana Modern Masters, London.

[21] Schon, D. (1983). *The Reflective Practitioner*. Basic Books, New York.

[22] Senge, P. (1990). *The Fifth Discipline*. Century Business Books, London.

[23] Turner, B. A. and Pidgeon, N. F. (1998). *Man-Made Disasters*, 2nd edn. Butterworth-Heinemann, Oxford.

[24] UK Parliament (2005). *Shipman Reports*. HMSO, London, https://webarchive.nationalarchives.gov.uk/20080820183409/http://www.the-shipman-inquiry.org.uk/finalreport.asp (Accessed on June 2019).

Chapter 8

Uncertainty — Prediction or Control?*

Abstract

Engineering practice involves decision-making under extreme uncertainty. The purpose of this chapter is to discuss the nature of uncertainty from first principles, to consider its various manifestations and to suggest ways in which we can improve. From first principles six kinds of uncertainty are identified under the headings of truth, trust, clarity, changeableness, incompleteness and risk. It is argued that truth and changeableness (together as one kind) with clarity and completeness are three central attributes of information and that trust is central to the making of decisions by 'experts'. They all are ingredients of the main task of managing and controlling risk. Reductionism has been successful for the analysis of hard systems but soft systems are governed by the behaviour of people which is so complex as to be hard to define and difficult to analyze. The emphasis in soft systems therefore is not on prediction but rather on managing a process to achieve desired outcomes based on dependable evidence which is context dependent. Processes are the way things behave in hard systems and what people do in soft systems.

---

*This chapter was originally published in *Int. J. Eng. under Uncertainty: Hazards Assessment and Mitigation*, 2009, 1, No. 1–2, 73–80.

Introduction

One common way of describing engineering practice is that it involves decision-making under extreme uncertainty [3]. Of course, this is such a wide statement that it also applies to many other forms of practical activity such as medicine. Indeed, as Weber [10] points out 'if there was no uncertainty, then we would never have to make a decision because we would be omniscient; we would always know what action to take...' Uncertainty analysis has tended to be dominated by the literature on decision-making and experiment design. Indeed, many authors categorize uncertainty as either epistemological or aleatory with cursory discussion before proceeding with their latest mathematical treatment of measurement errors or Bayesian probability.

Uncertainty in engineering practice is important and diverse. The purpose of this chapter is to discuss the nature of uncertainty from first principles, to consider its various manifestations and to suggest ways in which we can improve.

We will start by considering the meaning of the word certainty. Following Weber we will assume that certainty is not available to us by any other means than belief — usually of a religious kind — but we will not assume that this recognition implies that we cannot act with confidence to exercise proper duty of care. A survey of a number of dictionary definitions of certainty produced words such as truth, sureness, freedom from doubt or reservation, trustworthy, reliable or dependable, unquestionable, inevitable, indisputable, definite, particular or unambiguous. We can unpack these meanings by noting that there are three groups of kinds which I will later discuss in some detail. First is the age old philosophical and practical question of what we mean when we say something is *true* (e.g. truth, sureness) or not. In this I will include how we give 'meaning' or how we interpret what is expressed or indicated by a statement which may or may not be true. Second is the need to be able to *trust* what we say (freedom from doubt or reservation, trustworthy, reliable). Third is the level of *clarity* of what we say (definite, particular or unambiguous).

A similar survey of dictionary definitions of the word uncertainty tend to introduce other (opposite) terms like variable, erratic, unpredictable, indeterminate, sceptical, hesitancy, not assured or fully confident and ambivalent. It is also described as a quality of being with respect to duration, continuance or occurrence; a liability to chance or accident. There are eight headings for the word uncertainty in Roget's International Thesaurus. They are irregularity, changeableness, indistinctness, unsureness, equivocalness, irresolution, fickleness and precariousness.

In this chapter, I will distil these many further aspects into three more kinds of meaning which are *changeableness* (irregularity, variable, erratic), *incompleteness* (indeterminate) and *risk* (precariousness, unpredictability). I will show how these six kinds of uncertainty (truth, trust, clarity, changeableness, incompleteness and risk) are manifest in the processes of making engineering decisions and are orthogonal or independent. In other words, we can change one of them without affecting the others. I will argue that truth and changeableness (together as one kind) with clarity and completeness are three central attributes of information and that trust is central to the making of decisions by 'experts'. Finally, I will argue that all are ingredients of the main task of managing and controlling risk.

Ontology and Epistemology

Descartes famously said, "I think therefore I am". Heidegger used the term *da-sein* to express existence as a being-engaged in-the-world. Dias [6] sees Heidegger as having turned Descartes dictum around to argue 'I am in-the-world therefore I think', i.e. that existence and action is prior to knowing. It seems indisputable that being is prior to knowing since we are in the world, we are born, before we know it. Babies only become aware through their own consciousness. Prior to that the brain controls body functions and learning and then, as a baby develops, a conscious mind emerges with language that enables the kind of knowing that can be communicated and shared with others.

Heidegger used the activity of hammering in his analysis of 'being'. He wrote of the specific "handiness" or (ready-to-hand) of the hammer as we knock in a nail without being fully aware of what we are doing. But this is the kind of activity that is controlled through our subconscious minds and it is in that sense that 'we are because we act'. If we hit a thumb as we hammer our attention is immediately drawn as we become painfully aware of our hammering. A more modern example is the driving of a car when we operate almost without conscious thought until something requires it — such as a pedestrian stepping out into the road.

As our conscious thinking develops we learn. We learn from our personal experiences. We learn from others about accumulated human knowledge which is objective in the sense defined by Popper [6], i.e. that it exists outside any one individual. This knowledge may be true or false. We interpret what we learn through our own experience and all of its uncertainties and give it meaning. So it is in this sense that practice is prior to theory. However, doing and knowing leapfrog and build on each other in mutually reinforcing loops of learning. To know we must do and to do we must know.

All through history we humans have searched for certainty of knowing. Many find it through religion. But this is a certainty of revelation that is not easily shared with others. Uncertainty therefore arises both from what we do (how we behave) and from what we talk and write, i.e. how we communicate. Science is the search for certainty that can be shared. It is epistemological, i.e. derived from imperfect knowing, it is the study of the method, validity and scope of what we know. By contrast ontology is the study of existence or the nature of our being. Although doing or practical action is prior to our knowing about that action we can only understand and talk about our being through our knowing and we can only change and improve our acting through changing and improving our knowing. It follows that we can only study ontology through epistemology. We conclude therefore that all aspects of uncertainty are epistemological — even those concerning our existence.

Reductionism vs Systems Thinking

The idea that certainty is available is implicit in the scientific reductionist determinism that most of us were taught. Reductionism is the philosophy that any system can be reduced into parts and that if those parts are understood separately then you have an understanding of the whole. Determinism is the idea that all states are determined by antecedent states. Unfortunately, these concepts are rarely made explicit except to those who read philosophy.

Scientific reductionist determinism inevitably focuses on prediction since the normal demarcation between science and non-science is that scientific statements must be testable. In other words, we predict a state of affairs and then test to see if that comes about — if it does then we say it is true. Clearly that can only be done in within a certain context which is rarely explicitly recognized. To assert a universal truth we must resort to induction, i.e. we use a series of confirmatory particular cases to generalize. Alternatively, we evoke something like Popper's concept of the scientific method as conjecture and refutation in which falsifiability is the criteria to prove a theory wrong. So truth is always tentative and we can never be sure we have it.

The easiest and surest way to make something testable is to measure it. There is therefore a natural tendency to emphasize measurement and to focus on its uncertainties. Surveyors and experimental designers, for example, typically will consider human errors (e.g. mistakes in taking a reading), system errors (wrongly calibrated instruments) and random errors (repeated readings with all conditions as controlled as closely as possible). Surveyors will use many checks and techniques to reduce the chances of an error and will distribute those remaining in an attempt to obtain true readings. However, this begs the question of what constitutes a true reading. It is also a mistake to think that dealing with measurement errors is sufficient for dealing with uncertainty in all its many guises.

If it is not possible to test a statement directly we do the next best thing and test the source from which the information comes, i.e. we look for expertise. We test for internal inconsistencies.

Practicing engineers cannot afford to leave anything important out just because it is inconvenient or just too tricky to deal with. They do not have the scientist's luxury of selective inattention in order to isolate variables. They cannot define systems in a way which makes it easier — they have to define them as they are in total. They have to include elements of the context that makes the problem what it is even though they might prefer to standardize the process. For example, the replacement of a length of sewer pipe is much more complicated when there is a high voltage cable nearby. Approaches that ignore the human and social dimensions to engineering usually disappoint all concerned.

Practicing engineers therefore soon realize that their implicit reductionism (which is rarely stated explicitly and goes largely unrecognized) is inadequate. The reason is that practice involves making decisions under conditions of every kind of uncertainty. Whilst practicing engineers will always try to make statements that are testable there are many situations where this is not possible. The reductionist emphasis on prediction is only part of the systems thinking approach required to manage uncertain processes to successful outcomes [1].

Reductionism has been most successful for hard systems that are commonly said to be 'objective' in that they are supposed to be independent of the observer and hence the same for all of us. Reductionism has failed to make any impact on soft systems. Soft systems are, as we saw in Chapter 7, and as the name implies, systems which are hard to define — the edges are unclear. Generally soft systems are governed by the behaviour of people which is so complex as to be hard to define. One focus of systems thinking is the relationship between hard and soft systems. The emphasis in soft systems is not on prediction but rather on managing a process to achieve desired outcomes.

Of course within any system, hard or soft, there are statements that are or are not testable. Imagine that we disagree about some object in that I say it is heavy and ugly and you say it is light and beautiful. We can easily resolve the first disagreement because we have a way of measuring weight. Our disagreement about beauty however is more difficult since we have no agreed standardized way to test our statements — we cannot easily measure beauty. However, we can and do measure beauty, e.g. in beauty competitions or in ice-skating competitions. We know that different groups of experts may produce different answers — they are not repeatable — however that does not imply they are of no use. We can and do get useful information which enables choices and decisions to be made.

There is as yet no agreement about how we should measure soft issues — other than in specific circumstances such as political elections or competitions such as ice-skating championships. In any enquiry, the questions we ask will condition the answers we get. It is essential therefore that we first understand and communicate our purpose in asking the questions. The listening and observing and the collective understanding that follows can create added value. Thus, the measures are essential as long as the process and results are judged appropriate and fit for purpose.

Soft measurement is usually not as dependable as traditional hard measurement. For example, the opinions of people can swing for no obvious reason. However, this may not be an obstacle if we always start by defining the purpose of a measurement and we measure to address that purpose rather than trying to establish some 'true' value. If measurement, e.g. of satisfaction, is linked to an understanding of the reasons for the evaluation and that in turn is linked to action then we have a real opportunity to drive in improvement based on the measurement.

Successful professional people will not change unless they are presented with strong evidence to do so. Likewise, regulators and auditors will not be impressed with opinion and feelings — only by evidence and evidence is strongest when it is measured.

Processes are the way things behave in hard systems and what people do in soft systems. All designed hard systems have a function which is a role in a process. For example, a beam in a structure has the function of carrying the loads from the floor slab. A dam has the function of holding back the reservoir water. The steel and concrete of which the beam and the dam is made does not 'know' it has that function — it has no intentionality. The function is ascribed to a hard system by us, the people who own it, conceive it, design it, build it and use it. We are also the ones who decide when the hard system has failed and we decide the criteria of failure. Clearly some functions are obvious — others are less clear and unintended.

A measure of the quality of a hard system is its fitness for purpose as defined within a soft system. The purpose of a soft system is an ethical question. A measure of the quality of a soft system is its fitness for purpose in a hierarchy of human needs at the apex of which is human flourishing.

Blockley and Godfrey [1] have used the six attributes, 'who, what where, when, why and how' to characterize a process and to capture and record information for appropriate sharing with authorized other people. In this way each and every process is steered to success based on up-to-date integrated information.

I will now describe how truth and changeableness (together as one kind) with clarity and completeness are the three central attributes of information characterized using 'what, where, when and how'. Trust is central to people, characterized using 'who'. All are ingredients of evidence about risk which is the essential characteristics of a process which has to be steered to ultimate success.

Truth and Changeableness

How do we know the meaning of any statement and how do we judge its truth? Meaning derives from an interpretation of what is expressed and may or may not be true. I will use the only 'common sense' interpretation of truth which is credible to most practitioners, i.e. that a statement is true if it corresponds to the facts. Philosophically this definition is unfortunate as it leads to an infinite regress and we have to rely on a meta-system to establish the truth of the facts. The view that science gives us universal Truth (notice the big T) is discredited — yet its legacy lives on. So in practice how do we understand meaning and judge what is true? How dependable is that which we know? How do we give it value? These are central philosophical questions which have obvious practical implications for engineers.

Engineers, like all practitioners have a duty of care under the Law of Tort, to act responsibly. But what does that imply? To act responsibly engineers must always do what is reasonable. This means that they must be aware of the latest developments — particularly in science. They must know what should be reasonably known. They must hold the ethical values expected of them [4]. Values define quality. There is a continuum of quality in all things. This varies from the physical to the moral, from the idea of a good building to the idea of good conduct. The best way to understand that quality is through direct lived experience. Engineers have a legal and moral duty to deliver quality as a degree of excellence which is the state of having pre-eminence or having the highest value or the worth we give to the purposes we have both collectively and individually. Engineers have a legal duty to take responsibility to act reasonably. Ultimately the only test within the rule of law that stands credibility is that the peer group decides what is or what is not reasonable.

Within that legal context engineers require information which is sufficiently dependable for the complex decisions they have to make. Fortunately, universal Truth is a sufficient but not necessary condition for dependability. Newton's laws are not universally True in that propositions can be deduced from them that do not correspond with the facts for bodies that are travelling at velocities near the speed of light. However, Newton's laws are contingently true for almost all engineering decisions. Elastic theory is contingently true and dependable for stresses in steel from small strains but when strains are large plastic theory may be more appropriate. There are four conditions for a proposition to be dependable [3]. (1) A highly repeatable experiment can be set up to test it. (2) The resulting state is clearly definable and repeatable. (3) The value of the resulting state is measurable and repeatable. (4) The test is successful. Note that these conditions are to some degree sufficient but not necessary. Deficiencies in any of these conditions result in a loss of dependability.

It follows that when the state of the system is changeable we must look for patterns of behaviour. If we see regular patterns of change then we can hypothesize relationships which enable us to say that although the relationships are not True they are dependable for practical decision-making. One example is that the sun rises and sets every day — no one is likely to argue seriously that we should not assume that to be the case but there is no logical reason that it should. In his magnificent description of Popper's ideas, Magee [7] wrote, 'just because past futures have been like past pasts does not mean that future futures will look like future pasts'. In other words, we cannot take past regularities as the Truth but we can use them for dependable practical decision-making. As Mill [9] wrote, "There is no such thing as absolute certainty, but there is assurance sufficient for the purposes of human life".

We are familiar with hard systems analysis — it is the engineering science of structures, hydraulics, etc. These systems of structures and water pipes, etc. are described in terms of parameters such as span length, geometry, loading conditions, material properties, etc., which are precisely defined and measurable. Mathematical relationships are developed between these parameters by induction to describe system behaviour and to enable prediction. For example, many students of structural engineering will have shown the formula for the deflection of a simply supported elastic beam can be derived by testing various sizes of beams.

Now as we move away from the defined conditions under which a prediction has been tested then we may become increasingly concerned about the uncertainty in the relationship between the formula (which may be accurate in a laboratory test) and the actual situation. There are two types of uncertainty in this new situation — parameter uncertainty and model uncertainty.

Parameter uncertainty is the uncertainty about the values of the parameters that we may use in a calculation. It is the uncertainty covered by probabilistic reliability theory. Unfortunately, there is often an implied assumption that the system model is dependable and this is ignored by many analysts. If we have uncertainty in the systems model then we are uncertain of the relationship between a theoretical formulation and the actual system. This is a much more difficult problem and is one not satisfactorily treated in probabilistic reliability theory. If the concept and hence the equations are badly wrong then no safety factors can prevent failure. Engineers have to judge whether a model is appropriate or not and they do that by effectively testing when it has been used before and under what circumstances — i.e. they look for evidence that the model (in this case the formula) is dependable in a given context.

We therefore have, in practice, to treat the results of hard systems analysis as evidence in the soft systems decisions that have to be made.

The theories of statistics and probability are intended to deal with aleatory or random irregularities that depend on chance. Aleatory is dependence on contingent events or accidental causes of luck or chance. Randomness was defined by Popper as the lack of a specific pattern in information. More usually randomness is defined as proceeding without definite aim, reason or pattern, a situation where the various outcomes have an equal chance of happening — e.g. the tossing of a dice. Statistical theories assume that measured variables are random variables that show patterns of behaviour over large populations of trials. We now have available powerful techniques for dealing with irregularities identifiable over large numbers of trials. However, they break down when applied to specific cases. We can calculate the probability of the structural collapse of a building within a large class of similar buildings but we cannot be confident that the figure applies to the specific characteristics of a specific building. Theories of structural reliability can help to estimate the effects of the variability of specific parameters on a specific but 'notional' probability of failure. So, if the load on a beam and the yield strength are assumed to have defined probability density functions then we calculate a chance of failure for that specific structure. However, such calculations are very partial and reflect only a small part of the uncertainties (i.e. irregularities in parameter values) involved in assessing structural safety and reliability. The uncertainties in the basic theoretical models, when the parameters are precisely known, are very difficult to capture and are rarely included. When they are the models are relatively crude.

So in summary, truth and changeableness are basic attributes of information. Whether the information concerns measurable parameters or the patterns of relationships between those parameters which model our understanding of the behaviour of a system we must judge its dependability as evidence for the decisions we must make. Engineers can never be sure and never be free from doubt. They will always have reservations and must always question. They must never assume predictions are True or inevitable or indisputable. So questioning, testing, discussion and acting responsibly with a proper duty of care are the hallmarks of good practical decision-making. Any theoretical techniques of uncertainty analysis must always be supportive of this process.

Clarity

There are at least three types of lack of clarity in a piece of information. The first is that it is poorly constructed or difficult to understand — we will not consider that here since the only way to deal with it is to seek clarification. The important ways in which well-constructed information may lack clarity is through it being vague or ambiguous. However, we must immediately distinguish between vagueness and precision. If I say to you that my friend John is of medium height then I communicate something to you. My meaning is clearly communicated but it is not precise. It is vague, fuzzy or indistinct in the sense of not being precisely defined. A scientist may demand that I be more precise and demand that I state his height in metres. If I say he is between 1.80 and 1.82 m then I am more precise but less likely to be accurate and I may have to spend resources to get that information. In a practical situation I may be able to solve my problem without spending that resource. For example, if I want to buy him a shirt which comes in three sizes of small, medium and large then I have all the information I need.

It follows that precision is desirable but may not be necessary. Zadeh [11], an American systems scientist wrote, '... as the complexity of a system increases, our ability to make precise yet significant statements about its behaviour diminishes until a threshold is reached beyond which precision and significance become almost exclusive characteristics'. Zadeh maintained that the way human beings are able to summarize masses of information and then extract important items is because we think approximately. These ideas have led to important research into non-probabilistic methods of approximate reasoning [3].

Information is ambiguous if it has more than one meaning. Whilst the arts thrive on ambiguity, science and engineering require practitioners not to be equivocal and to be as clear and precise as possible. Therefore, it is part of an engineer's duty to explore any potential ambiguity and remove it wherever possible. There is no mathematics that can deal with ambiguity — it has to be managed by users. Ambivalence is similar to ambiguity — it is having mixed or contradictory feelings. It is a psychological uncertainty which can to a large extent depend on the individual. People's ability to absorb, process and react to information does vary with physical and mental attributes and this can generate uncertainty both in the individual and in others.

Incompleteness

The literature on incompleteness is dominated by Gödel's incompleteness theorems which state that formal systems rich enough to contain arithmetic contain undecidable propositions. A more general interpretation starts with the idea that all information is a reference to something else real or otherwise. It is not that thing but a description of it from a particular perspective. It is necessarily therefore often incomplete, i.e. not fully formed, unfinished and lacking in something. For example, a structural model of a bridge says nothing about its aesthetic value. A strict definition of incompleteness is that it is that which we do not know. However, in any model (scientific or otherwise) there will be some things that are purposely excluded (like the aesthetic value of a bridge in a structural model) and others that perhaps should be included if only we knew they existed.

Donald Rumsfeld, the one-time US Secretary of State for Defence was ridiculed in the media when he famously said, 'There are known knowns — these are things we know we know. There are known unknowns ... these are the things we do not know. But there are also unknown unknowns; these are the things we don't know we don't know'.

But Rumsfeld was right. Unless you are incredibly arrogant you must admit that there are many things that you do not know even exist. They may be known by others but not by you — you are completely unaware. Until lateral torsional buckling of beams was identified in the late 19th century it was an unknown to everyone — no one could recognize it even if presented with it in practice. When some of these things are identified then they become a known to someone but not to everyone. But perhaps, as an individual, you stop studying them before getting to the deepest level of understanding — you know they exist but that's all — they are a known unknown to you. We all have lots of these — quantum physics is one such for me.

Any lack of completeness can create unsureness. If information is wanting in some essential part, imperfect or indeed defective then it may not be fit for purpose. In summary, incompleteness has to be managed by understanding the context in which the information is dependable.

Orthogonality[a]

Figure 8.1 shows some characteristics of uncertainty such as ambiguity, confusion, contingency, indeterminacy and conflict as they emerge from mixes and interactions between the three basic FIR attributes as they are interpreted in any real-world context. Some theoreticians such as Hacking [14] and Huber [15] express the idea of unknown unknowns as ignorance — that usage is unfortunate. In a practical context, it can imply negligence as a lack of learning, being uneducated or uninformed or not properly qualified and hence not exercising a proper duty of care. Of course, ignorance of the law is no excuse but here we are referring to

[a]This section is extracted from Ref. [3].

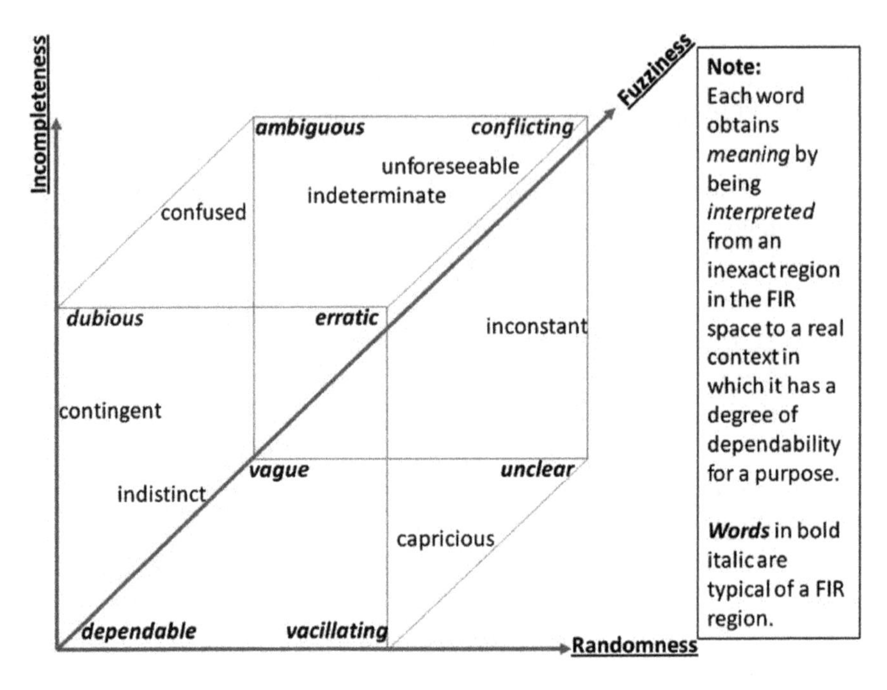

Figure 8.1. Some interpretations in the FIR space of uncertainty.

effects that no one properly qualified can reasonably foresee. Norton [16] attempts to capture ignorance by adopting the 'Principle of Indifference' and a 'Principle of the invariance of ignorance'. In this state, we have no grounds for preferring one proposition over any other and so we assign equal belief to an outcome and its negation symmetrically. Figure 8.1 indicates sets of possible 'mixes' for these expressions of uncertainty in natural language and Figure 8.2 illustrates the mapping from context to FIR space to an Italian Flag (see Chapter 9).

In natural language the subtlety of the meaning of each word will depend on an interpretation of the inexact region of an FIR mix in a context. I am referring to FIR as structural attributes rather than as attributes of interpretation and meaning. By structural I mean 'being constructed of — or the arrangement of parts of'. Meaning derives from an interpretation of the structural attributes in a context. Each usage of an interpretation may have a different mix of FIR depending on specific situations. So, for example, ambiguity emerges from interacting fuzziness and incompleteness that gives rise to a potential for more than one interpretation of the meaning of

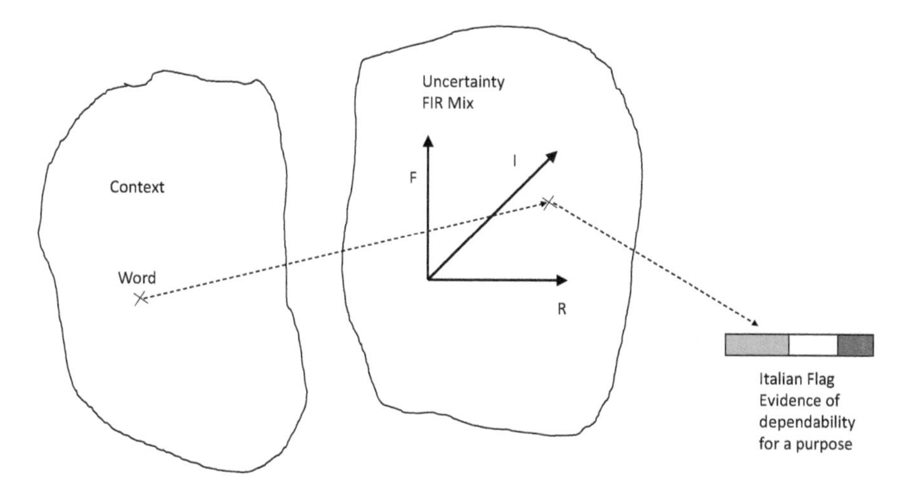

Figure 8.2. A mapping from context to FIR space to an Italian flag.

a statement. If someone ambiguously asserts that a proposition is 'somewhat true and somewhat false' they may mean that there is some rather vague evidence for and some evidence against. It is associated with dubiety where one hesitates to believe through mistrust due to incompleteness even when there is no explicit fuzziness as when a politician answers a question by leaving out important aspects of an issue. Erratic uncertainty emerges from interacting incompleteness and randomness so that interpretations are deviating, wandering and not fixed. The worst kind of uncertainty is where there is conflicting interpretations that are either not comparable or are incompatible or simply disagree. This can happen when all three parameters of uncertainty FIR occur simultaneously. The traditional way of handling these kinds of uncertainties is to demand clearer and more precise statements. Of course, this is a rational and wholly correct reaction and should be used wherever possible — but in complex problems that process if carried through insensitively may result in the loss of valuable evidence concerning poorly and incompletely understood phenomena.

Trust

A trustworthy person is someone you can depend on. There is no sense of irresolution, vacillation or fickleness. Uncertainty is created if a person is sceptical, hesitant, not assured or adequately confident. It will arise in a team if there is a lack of trust in anyone in a significant decision-making role. When people who have been trusted are found wanting then enormous damage ensues. There has been, in recent years, a loss of confidence in the idea of an 'expert'. A general questioning of authority, scandals in the City, corrupt lawyers who end up in jail, large-scale engineering failures have all contributed to the loss of faith in the 'expert'. The 'credit crunch' is a recent example. Professional people have to grapple with complex problems within complex systems and have to make judgements which lay people have to take on trust. Trust is a very fragile commodity which once broken is very difficult to rebuild. Once a few professionals are seen to be untrustworthy then unfortunately others will be dragged in and there is a reduction in the confidence in the profession as a whole. This loss of confidence leads to a dangerous 'relativism' where everyone's ideas are seen to be equal, no matter what the skills or expertise that person possesses. If major decisions are made this way because they are seen to be democratic, it is likely that the democratic society itself will be damaged. A healthy society has to be built on trust so that decisions are made by those best informed to do so and the rest of us feel able to trust them.

Targets are central to what Professor Michael E. McIntyre of the University of Cambridge has called our 'audit culture' [8]. He comments that this seems to have been driven by three very reasonable principles of fairness, objectivity and prudence. Fairness says that everyone must be treated in the same way. Objectivity requires that all assessments must be based on numerical measures. Prudence seems to suggest that no one can be trusted so we need to audit them. But who audits the auditors? How can the unmeasurables such as professional ideals and ethics, ambition, curiosity, enthusiasm, room for creativity, willingness to share be rewarded? He concludes, 'There is actually no alternative to reliance on trust — and to rebuilding trust where necessary. An advanced society will recognize this explicitly and live with the risks. It will use auditing resources in new and cost-effective ways by concentrating them not on trying to monitor and measure everything but, rather, on checks and balances against gross human failings' [7].

Risk

Precariousness is dependence on circumstances beyond one's control. It may imply a liability to the will of others or a reliance on insufficient or no foundation. It might mean dangerous or risky. Risk is about the future — it is the chance or likelihood of a specific hazard set in a specific context actually coming about. Risk is the chance of some state of affairs happening at some time in the future combined with the consequences that will follow.

Risks can only be understood in context. To identify, understand, monitor and change them we need dependable evidence. Evidence is information that helps us to come to a conclusion. It is the basis or reason for us to believe something — though we must hastily agree that, as individuals, we can quite easily believe something without evidence.

We have to take risks to explore the world — yet we all have our comfort zones — we have boundaries. To manage risks engineers need to be imaginative and creative with the foresight to think of everything that might happen and have contingency plans in place. Engineers know that unintended and unwanted things may happen.

Our collective scientific understanding and technical successes are impressive — sometimes we get carried away leading to high expectations that are sometimes not underpinned by a firm grasp of the risks. Failure can then seem like negligence — someone must be to blame. So how do engineers cope with human folly? They use their specialist knowledge and experience to plan very carefully, monitor everything that happens and change what they do to meet their objectives.

However, the issues are complex. The only way to get at a meaningful probability of failure is by collecting statistics over many years and over large populations. But, as noted earlier, these figures apply to all systems in general but none in particular. Prediction about any one specific thing is deeply uncertain. Practicing engineers manage risks not by making a scientific calculation of some mysterious number such as a probability of failure because they know that is insufficient. They look for dependable evidence that comes from scientific or mathematical reasoning, from information that has been tested, from opinions expressed by suitably qualified people through to legal testimonies in a court of law.

Engineers know that they have to be clear about purpose and objectives of what they are doing, monitor and decide on the basis of evidence. They have to collect evidence and use it to steer processes to success and manage the risks in a practically rigorous way.

One final aspect of risk is worth highlighting. When a small amount of damage to a system can cause disproportionate consequences that system is vulnerable even if the chances of that damage are very low. Risk consists of three equally important factors — the three C's of chance, consequences and context. Much of the risk literature focuses on chance. The much-neglected analysis of vulnerability concerns disproportionate consequences. Both are underpinned by insufficient explicit consideration of context.

All successful practicing engineers manage all three C's — often implicitly. It is hoped that this chapter might be a contribution to their improved explicit management.

Conclusions

(1) Uncertainty analysis has tended to be dominated by the literature on decision-making.
(2) Six key attributes of the evidence within decision processes are truth, trust, clarity, changeableness, incompleteness and risk.
(3) Evidence is context dependent.
(4) All aspects of uncertainty are epistemological — even those concerning our existence.
(5) Scientific reductionist determinism inevitably focuses on prediction.
(6) Practicing engineers cannot afford to leave anything important out just because it is inconvenient or just too tricky to deal with.
(7) Reductionism (which is rarely stated explicitly and goes largely unrecognised) is inadequate.
(8) Reductionism has been successful for the analysis of hard systems.
(9) Soft systems are governed by the behaviour of people which is so complex as to be hard to define and difficult to analyze.
(10) The emphasis in soft systems is not on prediction but rather on managing a process to achieve desired outcomes based on dependable evidence.
(11) Processes are the way things behave in hard systems and what people do in soft systems.

(12) The six attributes of 'who, what, where, when and how' characterize a process and to capture and record information for appropriate sharing with authorized other people. Engineers can never be sure and never be free from doubt. They will always have reservations and must always question.
(13) There is actually no alternative to reliance on trust — and to rebuilding trust where necessary.
(14) Risk consists of three equally important C's — chance, consequences and context.
(15) When a small amount of damage to a system can cause disproportionate consequences that system is vulnerable even if the chances of that damage are very low.

References

[1] Blockley, D. I. and Godfrey, P. S. (2000). *Doing It Differently*. Thomas Telford, London.
[2] Blockley, D. I. and Godfrey, P. S. (2004). Measuring judgements to improve performance. *Civ. Eng.*, 158, No. 3, 124–129.
[3] Blockley, D. I. (1980). *The Nature of Structural Design and Safety*. Ellis Horwood, Chichester.
[4] Blockley, D. I. (2005). Do ethics matter? *Struct. Eng.*, 83, No. 7, 27–31.
[5] Blockley, D. I. (2013). Analysing uncertainties: Towards comparing Bayesian and interval probabilities. *Mech. Syst. Signal Proc.*, 37, 30–42.
[6] Dias, W. P. S. (2007). Philosophical grounding and computational formalization for practice-based engineering knowledge. *Knowledge Based Syst.*, 20, No. 4, 382–387.
[7] Magee, B. (1978). *Popper*. Fontana Modern Masters, London.
[8] McIntyre, M. E. (2000). Audit, education, and Goodhart's law or, taking rigidity seriously, http://www.atm.damtp.cam.ac.uk/people/mem/papers/LHCE/dilnot-analysis.html.
[9] Mill, J. S. (1869). *On Liberty*. See https://en.wikisource.org/wiki/On_Liberty.
[10] Weber, J. A. (1998). Merging the metaphysical and epistemological aspects of uncertainty: A theoretical vision. *Pub. Admin. Management*, 3, No. 4, 34–52.
[11] Zadeh, L. (1973). Outline of a new approach to the analysis of complex systems and decision processes. *Trans. Syst. Man Cybernetics, IEEE*, 3, 28–44.

Part IV

Managing Risks to Find Resilience

Preamble

I first met Colin Brown in 1980 when he met me off the plane from Heathrow to Seattle. Colin was, at that time, a Full Professor of structural engineering at the University of Washington. He had invited me to lecture on what was my first visit to North America. Colin was to become my mentor and friend. Through him, I met Jim Yao at Purdue University and David Elms of the University of Canterbury in Christchurch, New Zealand, both of whom became long-term friends and collaborators.

Probability theory was and still is for many, the mathematics of uncertainty. Its use requires a different way of thinking about mathematics than the traditional mathematics most engineers were taught. Many engineers cannot see any reason why they should bother. On the other hand, most research workers in structural safety followed the line created by Alfred Freudenthal in 1947 when he introduced the use of probability theory into structural safety. His work was continued (amongst others) most notably by Gerhart Schueller, Masanobu Shinozuka, Allin Cornell, Niels Lind, Ove Ditlevesen and Rudiger Rackwitz. They all considered probability theory to be an essential mathematical tool and some (such as Niels Lind) as the only tool needed. Almost no one followed the line pointed out by Pugsley (Preamble, Part I) that there were more factors to consider than random variations in load and strength.

My experience told me that professional practice, such as engineering and medicine, often requires decisions to be made under highly uncertain conditions. When I met Colin Brown and later David Elms, Ian Munro and Paul Jowitt I realized that there were others who shared my discomfort about probability theory as the only theory of uncertainty. But, at that time we were all puzzled as to how probability and fuzzy sets 'fitted together'.

In 1990, I had a letter from Buenos Aires. It was from Professor Arturo Bignoli enthusing about fuzzy sets. Arturo later became a very good friend and who unfortunately died in 2018 aged 98. He had a wickedly teasing sense of humour which he expressed in what he calls 'Spanglish'. Arturo, born in 1920, lived all through the economic mess of Peron's Argentina and earned his living as a structural engineer and academic. He was Head of the Department of Civil Engineering at the

University of Buenos Aires and for a short time Rector of the new University Austral. Most intriguingly he was 'mad' keen on fuzzy sets. Every time we met Arturo had a small piece of paper on which he was constantly sketching out new ways of using fuzzy sets to help make engineering decisions. Through him I met Professor Alberto Bernardini of the University of Padua in Italy. Arturo, Alberto and I met to run seminars in our respective cities which Arturo dubbed 'The three Bs'. Alberto's interest is more mathematical than Arturo's much more practical take and I was somewhere in between — so we gelled in a way that is easy when it happens naturally but not easy to create when it has to be somewhat forced.

We (i.e. including Colin Brown, David Elms and 'The three Bs') all agreed that there was more to uncertainty than could be captured using probability theory. The questions were tough but it was clear that any attempts to bridge the gap between hard physical systems and soft people systems required us to examine the nature of uncertainty.

As a result I have come to see uncertainty as a gap of interdependence between a model and reality. We estimate it through a measure of truth-likeness. Unlike the philosophers of old who have searched and failed to find absolute certainty and absolute truth, I believe uncertainty to be profoundly uncertain. Absolute truth is only available through faith — by which I mean belief that is not based on demonstrable proof. One of the reasons is that truth depends on context. So, when we are faced with a decision to act based on our thinking our challenge is to understand and accept the nature of that uncertainty in context. The bridge we seek over the gap between thinking and doing has to be firmly founded on an understanding that all knowledge and information is interdependent contextual evidence.

I was sitting in my office one day in 1987 when the telephone rang. The voice at the other end was very Chinese and not easy to understand. The voice introduced himself as Weicheng Cui. Little did I know what was to flow from that brief conversation! He told me that he was studying for his PhD at Imperial College London but wanted to transfer to Bristol under my supervision — and he cogently listed the reasons why. It was difficult to argue with him because he clearly knew what he was talking about. I was immediately interested. Of course, I could not agree there and then — there are protocols to follow — but after some exchanges he did eventually transfer to Bristol. It was like taking on a whirlwind. Never in my experience before, or since, have I struggled to keep up with a student. Each week Weicheng produced pages of untidy highly mathematical notes. The notes were however comprehensive and logical so I did follow them and I did keep up. Weicheng obtained his PhD in Bristol in the minimum time possible — just over two years.

Weicheng's main contribution to my work in Bristol was to suggest a new measure of dependence between two sets. The measure ranges from mutual exclusion (and maximum perversity) through independence (no bias) to total dependence (where one set implies the other or in other words one set is contained in the other). Soon after Weicheng graduated Mauricio Sanchez-Silva came from Colombia to work with Colin Taylor and myself on a systems approach to earthquake vulnerability assessment. Mauricio, who is in 2019 a Full Professor at the Universidad de Los Andes, Bogotá, Colombia, took on Weicheng's methods and applied them. Something was missing however — we knew we were using only positive confirming evidence — we needed to include negative falsifying evidence if we were really following Popper's ideas (Chapter 1). Just after that colleagues John Davis secured a contract with the oil industry to exploit the methodology and then John worked with Colin Taylor and Jim Hall and John's PhD student Emad Marashi to produce a computer tool called Perimeta (Performance through Intelligent Management) to look at both condition monitoring and climate impact on the electric supply industry using a development of Weicheng's mathematics.

How is engineering safety best served by forensic science and engineering that pertains to legal processes? Professor Michael Furmston, a colleague at the University of Bristol, an academic lawyer and barrister wrote a chapter for a book which I edited called Engineering Safety [1]. He said that a critical question for a lawyer is 'who is trying to do what to whom?' The primary concern of the court is not an abstract inquiry into the causes of an incident but to establish who is to blame. Failure is essential to the growth of knowledge. As failure is exactly what engineers do not want, it is all the more essential that we learn lessons when it does happen. Technical reports are embedded in human and social systems and so forensic engineers must be sensitive to semantic subtleties regarding error, mistakes, accidents and disasters. **Learning Point No. 13** from this is that 'we need to distinguish how the courts allocate blame from the lessons learned'. I was invited by the editors to write an introductory paper for the first edition of the new journal of *Forensic Engineering of the Institution of Civil Engineers* [2]. Investigators into engineering failures are immediately faced with a task of disentangling technical and human factors. Forensic engineers really do need better methods for understanding hard and soft risks. As discussed earlier (Chapter 8) hard systems are the physical and technical matters traditionally dealt with by engineering science. Soft systems involve people and include matters traditionally dealt with by engineering management. In order to make improvements, engineers have to combine good-quality evidence from disparate sources, both technical and from wider issues.

In Chapter 9, we show how process models can be used to map the progress of projects. Disparate evidence can be measured and combined using interval probabilities drawn from Weicheng's mathematics as colourful 'Italian flag' indicators of risk. An Italian flag is associated with each process to indicate the level of dependability, based on all the information available at the time, that the process will be successful, which is to reach the stated objectives. Later I developed a new method of pairwise combinations and used it to calculate the flags through the entire process model with an example of the procurement of a building to illustrate the method. So, **Learning Point No. 14** is that 'the Italian flag can help judgements about incomplete knowledge'.

As I mentioned in Part III, Jitendra Agarwal came to Bristol to study for a PhD in 1991. Apart from the research on process described in Part II, Jitendra, Norman and I developed a general theory of vulnerability. Our first attempts focused entirely on structures and then only later did we attempt the more general systems approach set out in Chapter 13.

The first ideas for our theory of vulnerability came after I read a paper by David Elms called 'From a structure to a tree' published in the *Journal of Civil Engineering Systems* in 1983. The paper set out some potential civil engineering applications of mathematical graph theory. At the time I was teaching 1st-year civil engineers about the differences between (a) so-called just-stiff structures, (b) mechanisms and (c) redundant or statically indeterminate structures. David was writing about graphs which are networks of nodes interconnected by links. I saw that we could model structures as graphs. In the simple framed structures I was teaching at the time, the members would be links between joints at nodes. I realized that the links would sometimes carry more than one degree of freedom — but nevertheless we could use graph theory to investigate what happens when a link is lost — for any reason — probable or improbable. We decided that a system is vulnerable if it is susceptible to small damage cascading to disproportionate consequences as happened at Ronan Point in 1968 (Chapter 1) [3]. We set out to investigate the conditions in which this can occur using graph theory as our new basic tool.

I was aware at that time a number of structural engineering practitioners were concerned about robustness. Indeed, there was no formal theory of structural robustness. Structural designers who interpreted limit state design too literally could end up with structures that were optimum in a known context but not robust in a wider (and possibly unforeseen) context. Robustness was left as a matter of common sense and experience. For example, just because a calculation might demonstrate that a steel plate of only a few millimetres was adequate in a particular situation the experienced designer relying on common sense would specify a thicker plate that would survive the fabrication and construction processes it would have to go through before working in the final structure. My concern about robustness was more theoretical — an inappropriate use of optimization ignoring context. Many theoretically inclined structural engineers were keen on newly powerful optimization methods based on linear and non-linear programming. I was intuitively opposed. I argued that if you optimize a system against a set of criteria then that is fine *as long as the context does not change*. However, when the context does change, as it invariably does in practical matters, there is suddenly a potential for your structure to lack robustness. One simple example is a deep I beam. When the compression flange is restrained laterally the optimization criteria involve only beam bending. If the constraints are removed then the beam becomes vulnerable to lateral torsional buckling. So, it is in this sense vulnerability is sufficient for a lack of robustness.

As we developed our theory we focused on the form of the structure under any possible loading condition — something we thought was powerful and incisive. The problem we faced is that few engineers who read our work could see how we could 'get-away' with not looking at loading conditions. They had difficulty with our basic approach because we were looking at the form of the structure rather than the effects of a load on the structure. As a consequence many rejected our theory as being of limited use. At about the same time we realized we needed to relate these ideas to resilience. Resilience is considered as the ability of a system to withstand or recover quickly from difficult conditions. It is not a simple property like a safety factor or probability of failure; rather it emerges from the interactions between sub-processes. Resilience engineering represents a new way of thinking which does not regard safety as something a system has (a property) rather than something a system does (a performance). One measure of unintended consequences is the number of 'surprises' experienced.

Our previous work on uncertainty and failure helped us to understand that complex infrastructure systems may contain new risks through interdependencies that may not be fully understood. Systems thinking tells us that we cannot always predict the total behaviour of a complex system from the performance of its interdependent parts. Complex systems are by their very nature difficult to manage. They are typically composed of many interconnected parts, are difficult to understand, explain or analyze and hence model. They are different from systems that are simply complicated because the latter may have many parts but the performance of the whole can be predicted by modelling the performance of the parts. Players in a complex system have to expect and devise ways of dealing with unexpected and unintended consequences.

We decided to define a system as vulnerable if it is susceptible to damage or perturbation where small damage or perturbation leads or cascades to disproportionate consequences. We decided that robustness is necessary but not sufficient for resilience and resilience in turn is necessary but not sufficient for sustainability. All are handled by systemically managing risks.

Designing for foreseeable risks is a challenge but accounting for risks which are difficult or even impossible to foresee — such as those arising from complex interdependent processes — poses a far greater challenge. This book argues that civil engineers need a way of addressing such low-chance but potentially high-impact risks if they are to deliver truly resilient infrastructure systems. They need to cultivate a wisdom to admit what they genuinely do not know, and to develop processes to manage emerging unforeseeable consequences. **Learning Point No. 15** is that 'complex systems may contain new risks through unknown interdependencies'.

References

[1] Blockley, D. I. (1992). *Engineering Safety*. McGraw Hill, London.
[2] Blockley, D. I. (2011). Engineering Safety. *Proc. Instn. Civ. Engrs. Forensic Eng.*, 164, No. FE1, 7–13.
[3] Ministry of Housing and Local Government (1968). *Collapse of Flats at Ronan Point, Canning Town*. HMSO, London.

Chapter 9

Managing Risks to Structures*

Abstract

Structural engineering has changed markedly over the last decades, creating new challenges and new opportunities. Consequently, structural engineers are widening their thinking from just technical issues to the effects of other matters on the risks to their structures. In particular, there is a need to find better methods for integrating hard and soft risks. Hard systems are physical and technical matters traditionally dealt with by engineering science. Soft systems involve people and include matters traditionally dealt with by engineering management. In order to make improvements, engineers have to combine good-quality evidence from disparate sources, both technical and from wider issues. This chapter demonstrates how disparate evidence can be measured and combined using interval probabilities drawn as colourful 'Italian flag' indicators of risk. Process models are used to map the progress of projects. An Italian flag is associated with each process to indicate the level of dependability, based on all the information available at the time, that the process will be successful, which is to reach the stated objectives. A new method of pairwise combinations is described and used to calculate the flags through the entire process model. An example of the procurement of a building is used to illustrate the method.

* This chapter was originally published in *Proc. Inst. Civ. Engrs. — Struct. Build.*, 2008, 161, No. 4, 231–237.

Introduction

Structural engineering has changed markedly over the last decades. Not only are new powerful computational tools available, but there are also different challenges such as terrorist attack and the impact of climate change. Consequently, there is a much greater awareness of the need to deal explicitly with risk in all aspects of life. A key difficulty is how structural engineer decision makers integrate information from many disparate sources to manage the risks to structures; this is the central topic of this chapter.

Integrating Hard Physical Risks with Soft People Risks

In many projects, accountants manage the known financial risks well, the engineers manage the known technological risks well, the safety specialists manage the known health and safety risks well, the quality managers manage the known processes well and so on. Major problems, however, even in successful companies, seem to arise in the gaps between these specialisms, resulting in unknown and unintended complications such as cost and time overruns and consequent quality problems. Even within specialisms, however, a range of techniques may be used to assess different aspects of risk which are difficult to integrate. The problem to be addressed in this chapter is how to facilitate good balanced decisions to manage all of the risks and hence minimize unintended harmful consequences: in short to improve the ability to do 'joined-up' thinking across a fragmented industry.

Hard systems are physical systems that are commonly said to be 'objective' in that they are supposed to be independent of the observer and hence the same for all of us (Chapter 8). Hard systems are the topic of traditional engineering science. Soft systems are, as the name implies, systems that are difficult to define — the edges are unclear. Generally, soft systems are governed by the behaviour of people, which can be complex. Soft systems are the topic of traditional engineering management. The emphasis in soft systems is not therefore on prediction, but rather on managing a process to achieve desired outcomes.

Processes are the way things behave in hard systems and what people do in soft systems. All designed hard systems have a function. For example, a beam in a structure has the function of carrying the loads from the floor slab. A dam has the function of holding back the reservoir water. The steel and concrete of which the beam and the dam are made does not 'know' it has that function — it has no intentionality. The function is ascribed to a hard system by the people who own it, conceive it, design it, build it and use it. This function can therefore be perceived as a role in a process. Part of defining that role is to decide the criteria of failure. This is done by different people from different points of view and may be contentious. Clearly some functions are obvious; others are less clear and unintended. For example, a bridge designed to carry road traffic was almost certainly not designed to be used as a shelter by homeless people. In one case the cost of repair to concrete damaged by the fires lit by homeless people to keep warm under a bridge was substantial.

In summary all designed hard system processes are embedded in one or more soft system processes.

Process Models

A process model is used as the integrating structure on which everything else is built [1]. There are many views of what constitutes a process. Most people use the word to distinguish the form of the 'doing of an activity' from the content which is the output, for example, a product. Here a new way of thinking about a process is applied, which does not separate form and content. This new concept of what constitutes a process is much richer than an input transformed into an output, a recipe, a Gantt Chart, a network or a flowchart, although it can be simplified down to each of these if required. In this methodology each process is regarded as a holon, that is it is both a whole and a part at one and the same time. It also has emergent properties [1], from systems theory, that arise from the complex interaction of the parts. The relationships between processes are captured in a process map and then each individual process is used as a 'peg' on which to attach all other data and information. The process map therefore sets out the basic structure of a project and all of the data associated with it. The traditional separation of process from product is lost because, by this view, a process can be the 'doing of an activity' or the 'performing of a physical system' — the structuring of the information is the same for both.

In order to use this idea, the map of the interconnections has to show how processes relate to each other. This is the purpose of a 'project progress map' (PPM): it enables various data such as risk registers, structural calculations, project progress measures and so on to be integrated. As stated above, the processes in the PPM form a central spine or skeletal structure on which all these data and attributes are attached.

The traditional view is that a product is the output of a process. It is useful to keep these two ideas distinct because it defines what clients perceive they are buying. However, it is important to understand that in the proposed methodology, because products do things and exist through time, products are also processes seen in a wider perspective.

The process holons are arranged hierarchically in layers. Initially the whole system is described as one process: the 'top' process and the top layer. Obviously to achieve success of this top process many sub-processes have to be successful and these are the second layer. For the sub-processes to be successful some sub-sub-processes must be successful and this continues down to a level of detail as appropriate for the particular system. This is not a reductionist approach because the many interactions between processes at the same layer are recognized and modelled as far as they are understood. For example, time relationships can be included using a standard critical path network analysis.

In this methodology, each process is monitored and adjusted to keep it on a path to success. Success is defined by a set of precisely stated objectives. Every process should have a process owner. The process owner is the person who is responsible for delivering success of that particular process. Parameters and performance indicators of the process are monitored to keep them within required bounds and to provide evidence of trends that need remedial action. In this way unintended consequences can be identified early. Thus, at any given time during a process evidence is available from past and present performance and predictive analyses about the future. The words 'who, what, where, when, why and how' are formally used to capture and record information for appropriate sharing with authorized other people. In this way each and every process is steered to success based on up-to-date integrated information.

Measuring Evidence of Performance Using an Italian Flag

A key idea of the methodology is to find evidence that any given process is moving towards success and there is no build-up of difficulties that might bring about failure. Opportunities for improvement should also be located. The evidence will come from many sources of various types and there is a need to collect it, digest it, interpret it, learn from it and make decisions using it.

In order to have a measure of evidence that can be used in soft as well as hard problems it is necessary to assess, quite separately, the evidence in favour and the evidence against the proposition that a process is heading for success. This can be conceived of in several ways. One way is as a vote by a group. Another is as an individual responsible judgement — a kind of internal vote. A scale of [0, 1] is used as a measure of evidence in favour and evidence against.

The degree of evidence that a process is heading for success is coloured in green, as shown in Figure 9.1 Evidence against is also assessed on a scale [0, 1] and is coloured in red starting from 1 and working back to zero. The difference in the middle is white and represents incompleteness — the extent to which we do not know. The three colours together make the Italian flag.

An all-green flag means that there is complete evidence for and no evidence against (no red). An all-red flag means that there is complete evidence against and no evidence for (no green). An all-white flag means there is no green evidence for and no red evidence against and so we really 'do not know' or indeed have no view.

Figure 9.2 shows a simple model of a process with two sub-process holons, the successes of which are together necessary and sufficient for the success of the top process. The players who own each process associate an Italian flag with that process. The flag represents their view, based

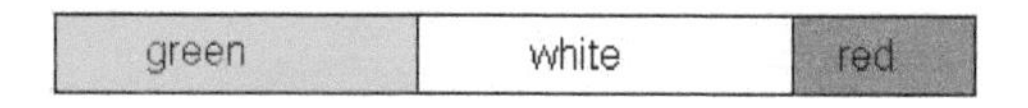

Figure 9.1. An Italian flag.

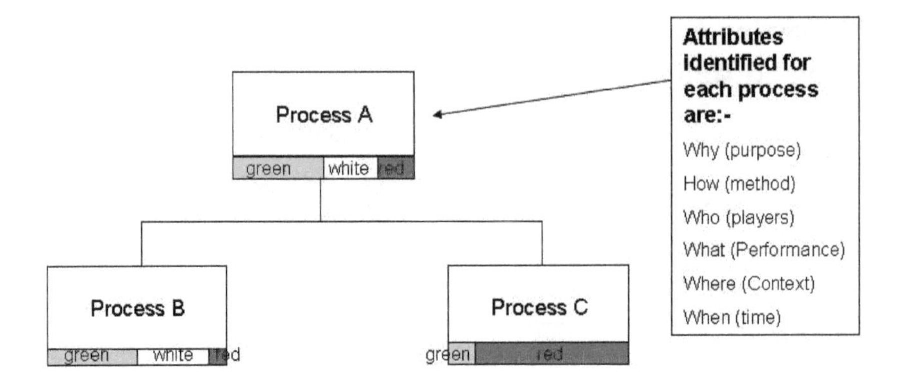

Figure 9.2. A map of three simple processes.

Note: The Italian Flag of evidence for Process C indicates that it is very likely to fail. Unfortunately the evidence available to the owner of process A indicates that it will be successful. If process C is necessary for Process A then the two process owners need to sort out the reasons for their differing perceptions.

on evidence, that their process will be successful. Clearly there is something wrong in Figure 9.2 since the flags are inconsistent in a rather blatant way for the purpose of the example. This means that when the process owners realize this inconsistency, they can discuss the reasons for it and decide on adjustments or on what needs to be done to improve the chances of success.

Until recently the flags were used entirely qualitatively as described by Blockley and Godfrey [1]. Now a new algorithm developed for the purpose of propagating the flags mathematically is described. Certain constraints and assumptions need to be appreciated, however, before the method is used.

The Mathematics of Italian Flags[a]

Mathematically an Italian flag is an interval probability. That just means that the probability of an event or proposition E has a lower bound El and an upper bound Eu.

[a]NB: The next two sections may be skipped without loss if you do not want to delve into detailed mathematics.

So, $p(E)$ is an interval number

$$P(E) = [El, Eu] = [g, (1 - r)].$$

We use g to represent the green part of the flag which colours the interval on the scale [0, 1] from 0 to g. Likewise r is the red part of the flag from $(1 - r)$ to 1. The white part of the flag is $w = 1 - g - r$.

First, we will calculate the Italian flag for a proposition H based on the evidence expressed as a flag for a single piece of evidence E. We define H as the proposition 'process H will be successful'. E is a sub-process of H at the next layer down. The term $p(H)$ is the probability of H and a measure of the dependability that the process H will be successful. The term $p(E)$ is the probability of E and is a measure of the dependability that E will be successful. The term $p(-H)$ is a measure of the dependability that the process H will fail and $p(-E)$ is a measure of the dependability that E will fail.

The total probability theorem tells us that

$$p(H) = p(H/E) \cdot p(E) + (1 - p(-H/-E)) \cdot p(-E),$$

where $p(H/E)$ is the probability of H given E and a measure of the dependability that H will be successful given that E is successful. Likewise, $p(-H/-E)$ is the probability of *not* H given *not* E and a measure of the dependability that H will fail given that E has failed. The values of $p(E)$ and $p(-E)$ are obtained not only through data from measurements, judgements or calculations but also may be propagated from calculations on sub-processes passed up through the process hierarchy. The values $p(H/E)$ and $p(-H/-E)$ are input by the user. The first measure $p(H/E)$ represents a warrant or justification for an action or a belief in the success of H if E is totally successful — it is the degree to which success in E provides positive support (assurance, affirmation) for the success of H. In a similar way $p(-H/-E)$ represents a warrant or justification for an action or a belief in the failure of H if E totally fails: it is the degree to which the failure of E provides negative support (testability and falsification) for the failure of H.

So, $p(H/E)$ is called the positive or sufficient support and $p(-H/-E)$ is called the negative or necessary support. They are judged by asking the following questions for each sub-process separately regardless of all other sub-processes:

(a) **Positive support:** if E succeeds what are the chances that H will succeed regardless of the evidence for other sub-processes?
(b) **Negative support:** if E fails what are the chances that H will also fail regardless of the evidence for other sub-processes?

Figure 9.3 gives guidance on the implications of various choices.

$$\text{Note that } p(H/-E) = 1 - p(-H/-E).$$

We express the probability of H as

$$p(H) = [Hl, Hu] \text{ and}$$
$$p(H/E) = [Sn(H/E), Sp(H/E)] = s = [sl, su],$$

so, $Sn(H/E)$ is a lower bound on $p(H/E)$ which is more simply expressed as sl and $Sp(H/E)$ is an upper bound on $p(H/E)$ which is more simply expressed as su and s is the interval number $[sl, su]$.

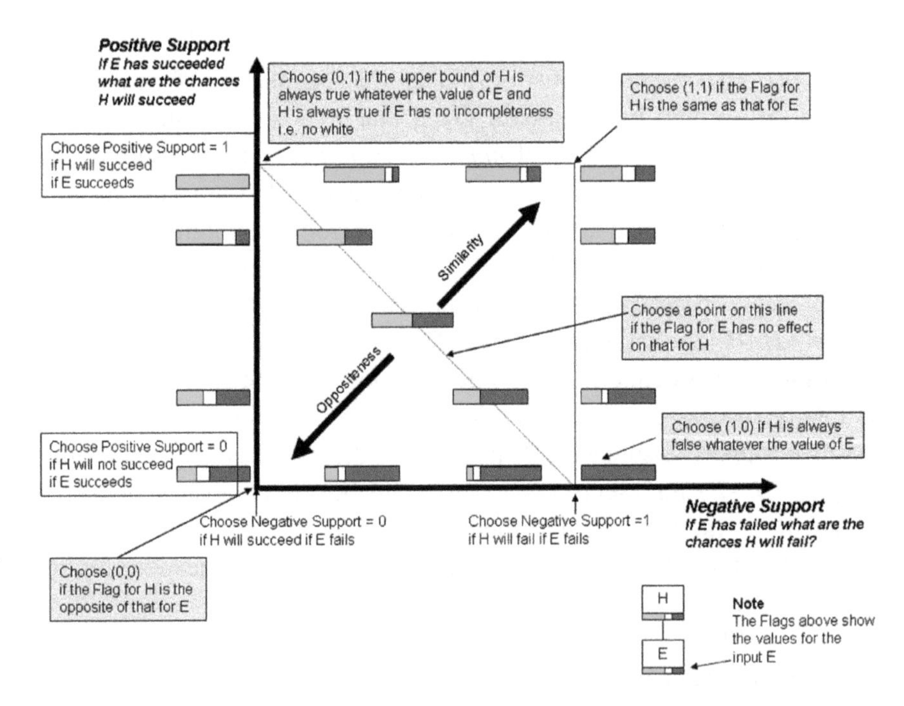

Figure 9.3. How to choose support values for the warrant of E on H.

In a similar way if the probability of *not H* is $p(-H)$ then

$$p(-H/-E) = [Sn(-H/-E), Sp(-H/-E)] = n = [nl, nu],$$

so, $Sn(-H/-E)$ is a lower bound on $p(-H/-E)$ which is nl and $Sp(-H/-E)$ is an upper bound on $p(-H/-E)$ which is nu and n is the interval number $[nl, nu]$.

From probability theory,

$$p(-H/E) = 1 - p(H/E) = 1 - s = [Sn(-H/E), Sp(-H/E)] = [1 - su, 1 - sl]$$

and

$$p(H/-E) = 1 - p(-H/-E) = 1 - n = [Sn(H/-E), Sp(H/-E)]$$
$$= [1 - nu, 1 - nl].$$

Then we calculate $p(H)$ using a variation on the total probability theorem as derived by Marashi [2] as follows:

$$p(H) = p(H/E)g + p(H/-E)r + (p(H/E)\, p(H/-E))w,$$

$$p(-H) = p(-H/E)g + p(-H/-E)r + p(-H/E)\, p(-H/-E))w.$$

The reason for using this variation is because a simple interval version of the total probability theorem would violate some of the basic probability constraints. A numerical example to follow below will illustrate how this is so. To avoid this in Marashi's variation we operate separately on the three intervals: g, w and r. The conditional probabilities $p(H/E)$ and $p(-H/E)$ operate on the g and $p(H/-E)$ and $p(-H/-E)$ operate on the r as standard but the white w is treated differently. We assume it to be a quantity statistically independent of both conditionals. In other words, it has least bias — so we multiply them together.

We find the upper bound on H using the upper bound on $p(H/E)$ that is su operating on g. Likewise, we find the upper bound on $p(H/-E)$ but this time by using $(1 - nl)$ operating on r:

$$Hu = Sp(H) = su\, g + (1 - nl)r + (1 - (1 - su)nl)w.$$

We calculate Hl in a similar way but now we use the opposite bounds on s and n, that is sl and nu to calculate an upper bound on $-H$ before obtaining the lower bound on H. Thus, the maximum of $-H$, that is, $-Hu$, is calculated and we subtract it from 1:

$$Hl = Sl(H) = 1 - Sp(-H) = 1 - \{(1 - sl)\, g + nu\, r + (1 - sl(1 - nu))w\}.$$

Note we use the upper bound is on $p(-H/E)$ which is $(1 - sl)$, to maximize the green on $-H$ so that in turn we minimize the green on H. Likewise we use the upper bound on $p(-H/-E)$ or nu to minimize the red on $-H$ and hence maximize the red on H.

So, in summary,

$$p(H) = [Sn(H), Sp(H)] = [Hl, Hu],$$

where

$$Hl = 1 - \{(1 - sl)\, g + nu\, r + (1 - sl(1 - nu))\, w\},$$

$$Hu = su\, g + (1 - nl)\, r + (1 - (1 - su)nl)w.$$

We can illustrate the results with a numerical example:

Assume that $p(E) = [0.3, 0.7]$, that is $g = 0.3$, $r = 0.3$, $w = 0.4$ and $[sl, su] = [0.4, 0.6]$, $[nl, nu] = [0.4, 0.8]$.

We cannot use a simple interval calculus because it produces inconsistent results as follows:

$$p(H) = [0.4, 0.6] \times [0.3, 0.7] + [1 - 0.8, 1 - 0.4] \times [0.3, 0.7] = [0.18, 0.84]$$

and hence $p(-H) = [1 - 0.84, 1 - 0.18] = [0.16, 0.82]$.

Whereas, alternatively $p(-H) = [0.4, 0.6] \times [0.3, 0.7] + [0.4, 0.8] \times [0.3, 0.7] = [0.24, 0.98]$ and hence $p(H) = [0.02, 0.76]$ and the two sets of results are incompatible.

These inconsistencies are removed using the formulae derived earlier:

$$Hl = 1 - \{(1 - 0.4) \times 0.3 + 0.8 \times 0.3 + (1 - 0.4 \times (1 - 0.8)) \times 0.4\}$$
$$= 1 - \{0.788\} = 0.212,$$

$$Hu = 0.6 \times 0.3 + (1 - 0.4) \times 0.3 + (1 - (1 - 0.6) \times 0.4) \times 0.4$$
$$= 0.18 + 0.18 + (1 - 0.16) \times 0.4 = 0.36 + 0.84 \times 0.4 = 0.696,$$

and so $p(H) = [0.212, 0.696]$.

Propagation of Evidence by Pairwise Combination

Now, we can set about calculating the Italian flag of evidence $p(H)$ for a process H, which has a number of sub-processes $E1$, $E2$, $E3$ and so on with associated supporting evidence flags of $p(E1)$, $p(E2)$, $p(E3)$ and so on.

A sub-process may be declared as jointly sufficient or jointly necessary. If a sub-process is declared as being jointly sufficient then the flag for any process calculated using that sub-process will not have a lower bound less than the lower bound for this sub-process. This means that, in this condition, the green must be at least as big as the green for the sub-process. Likewise, if a sub-process is declared as being jointly necessary then the flag for any process calculated using that sub-process will not have an upper bound higher than the upper bound for that sub-process. This means that, in this condition, the red must be at least as big as the red for the sub-process.

In order to deal with the complexity of multiple dependencies between sub-processes we use a pairwise combination method. We first calculate the evidence flag for a process based on the flag for each sub-process flag separately using the total probability theorem as described above. Any requirements of joint sufficiency or necessity are not included at this stage. This therefore results in a series of separate estimates of $p(H)$ based on Ei, referred to as $p1(H)$, $p2(H)$, $p3(H)$.

These separate estimates are then combined in pairs. We use dependency values *rho* to express the dependency between each pair [2]. The values *rho* are a measure of the overlap or commonality between processes. They are used to reduce the likelihood of 'double counting'. The dependency values range between total dependency, through independence to mutual exclusion and maximum perversity and are chosen by the user.

The first modelling assumption as set out in Figure 9.4 is made by assuming that the dependency between *pi*(*H*) and *pj*(*H*) is as between *p*(*Ei*) and *p*(*Ej*).

Any conditions of joint sufficiency or necessity are included by checking the sizes of the green and red. There can be conflict between evidence when there is some support for *H* and also for −*H*. We automatically redistribute the conflict to the white interval. However, if the inputs are logically inconsistent then we get inadmissible conflict. When this happens a section (*E*2 & *H*2) appears in the Italian flag, indicating its relative size. The only way to resolve this is by changing the input values.

So now we have a series of results *pij*(*H*) and they have to be combined into a single answer. We use a second modelling assumption with two alternative solutions depending on the purpose of the user. The first is to assume that the results *pij*(*H*) are all totally dependent. In this

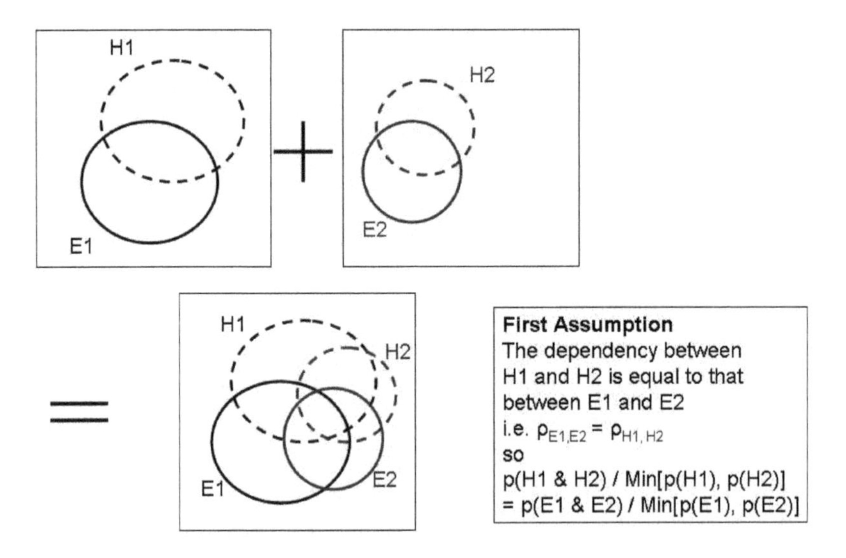

Figure 9.4. Nature of first assumption.

worst-case scenario, the minimum green (smallest positive evidence) and minimum red (smallest negative evidence) are to produce maximum white (maximum do not know). The second alternative is less severe and takes the average of the greens and the average of the reds. These indicate the total effect of the positive and negative support for H. Any requirements of joint sufficiency or necessity are included at this stage.

The choice of which way to combine the $pij(H)$ will be made according to the needs of the process players. Whilst these choices may, at first sight, seem *ad hoc*, it is important to remember that the purpose of this methodology is not to provide accurate predictions of the future — rather it is to enable the players to explore and understand the sources of uncertainty and hence to make appropriate decisions to improve the chances of attaining success. Both of these assumptions can reveal different particular insights.

A Structural Example: Procuring a New Building

What are the risks to the procurement of a new building? We can start with an overview for all of the players in the process. Traditionally particular players, such as structural engineers, may not be appointed to be concerned with high-level processes. However, in understanding their role in an integrated team the overview top process sets the critical context for the success of their work.

Following the process methodology described by Blockley and Godfrey [1] the first task is to create a process model. An example is shown in Figure 9.5. The reader can get a glimpse of an overview of the totality of the interdependent processes in play from this figure even if each box is too small to read in detail. The figure does not include all of the detailed processes necessary to procure a building because of the limitations of space and so necessarily the model is incomplete.

We can however draw some overview conclusions. The first point to note is that the model is hierarchical. The top process is 'procuring a new building'. It represents the whole of the project and recognizes the complexity of all of its emergent properties. Its associated Italian flag is an indicator of the total risk to the success of the whole project. The second point is that the flag will change through time as evidence about the risks

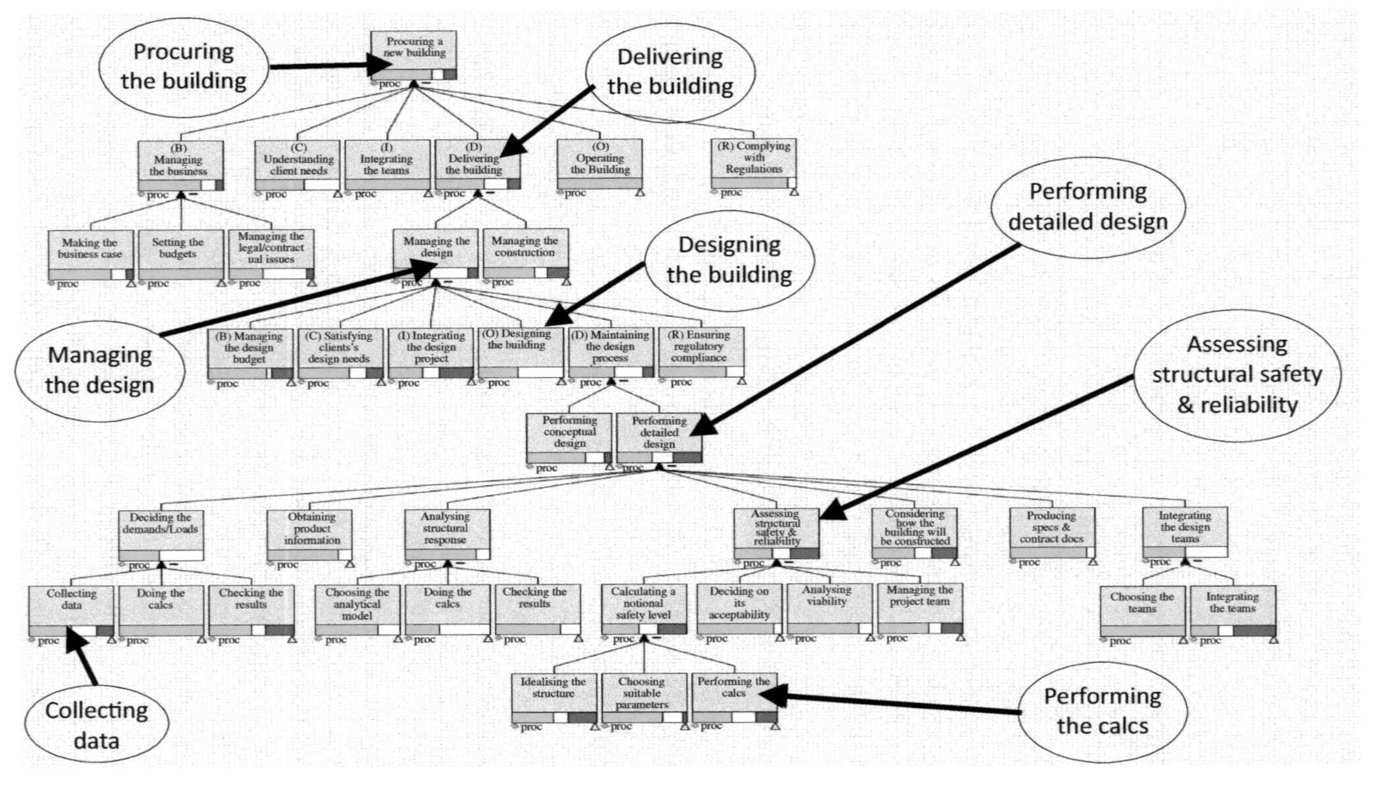

Figure 9.5. Procuring a building.

change. Third, one of the sub-processes of the top process is 'Delivering the building' and this in turn has a sub-process of 'Managing the design'. Only in the fourth layer do we find 'Designing the building'.

Fourth, the model is developed by identifying at the next level down all of the sub-processes which together are necessary and sufficient for the process above to succeed. Then by looking at each next layer process in turn further sub-processes are identified. Again they are the processes, the successes of which are together necessary and sufficient for the success of a specific process in the layer above. The building of the model is then repeated for further levels to an appropriate level of detail such as 'performing the calculations' in Figure 9.5.

As stated earlier formally the flag illustrates an interval probability measure of the dependability that a process will be successful based on the available evidence from the identified sub-processes. At and before the start of the project if there is no evidence at all about the risks then the flag will be completely white. The Italian flags for the bottom processes in particular (or all processes in general) can either be input directly by the relevant process owner or derived from a performance indicator using a value function as described by Hall *et al.* [3]. The flags for the higher process, right to the very top, are then calculated using the pairwise algorithm described earlier so that the effects of the lower flags on the higher flags become evident. At the same time, process owners will have their own interpretations of the evidence available to them and hence their own versions of the flags. Any important discrepancies as illustrated by Figure 9.2 are highlighted and appropriate decisions taken in a timely manner.

The system is dynamic and constantly changing as new evidence arises and process owners change their flags. Of course, all of this means that success for each process has to be defined and updated precisely. The system can also be a rich source of other project information with the advantage that it is structured in an integrated way. This is achieved by attaching information to each process. Some of that information will emerge from lower processes under the labels 'who, what, where, when and how'. However, perhaps the most important attribute of all is that each process has an 'owner' who monitors the flag and is responsible for identifying actions required to keep the process on track. This, of course, may well involve negotiating with others about changes they may need to make, since processes are interdependent.

Note that Figure 9.5 includes all processes — both soft and hard. In other words, the model is not restricted to the physical building but includes all information of all kinds. The pedigree of the information clearly varies enormously and that is catered for through the input data. Thus, if a difficult judgement, which is necessarily imprecise, is put alongside information from a numerical measurement or calculation then this can be allowed for by a suitable adjustment of the relationships between the measurement and the flag.

The structuring of the information can be implemented on a project intranet so that the model would be available to all participants in a project (with appropriate security safeguards) with read only or write access to their 'pieces of the action'. Process owners would be responsible for feeding in their latest data. After an initial input of contact details and other relatively stable information the major changes may be values of state variables (e.g. hard systems: movement of a foundation or dynamic responses of a bridge deck), key performance indicators (KPIs) (e.g. soft systems — judgements about team behaviour or loss of key personnel) and/or judgements about the Italian flags. The inputs could include the results of very detailed calculations such as finite element analysis or even notional probabilities of failure. Other examples are detailed design and construction activities together with judgements and notes about the context and dependability of any calculations. In this way they can be interpreted and integrated into the total picture as dependably as possible.

A quick scan over all the flags will enable the spotting of possibly previously hidden interactions that could cause problems. In this way, decisions are taken by everyone involved in full awareness of the impact of other decision makers on total project success.

Conclusions

(1) The need to integrate different risks from different types of information has been highlighted. It has been argued that structural

engineers have to understand and deal with the risks to their projects from the complex interactions between technical and non-technical issues.

(2) The use of process maps to connect data, information, people and functions and provide routes by which change can be managed has been described and illustrated with an example of the procuring of a building.

(3) The clear distinction between hard and soft systems has been made. It has been shown that they can be integrated by modelling them both as an interacting set of process holons.

(4) Italian flags based on interval probability theory are a practical way of representing the dependability of evidence in a process model. Some new mathematics of interval probability has been presented, which enables the interval values to be propagated up through the hierarchy of a process model.

(5) The Italian flag can be used as part of a social process to improve understanding of, and judgements associated with, incompleteness of knowledge.

References

[1] Blockley, D. I. and Godfrey, P. S. (2000). *Doing It Differently*. Thomas Telford, London.

[2] Marashi, E. (2006). *Managing Discourse and Uncertainty for Decision-Making in Civil and Infrastructure Engineering Systems*. PhD thesis, University of Bristol, UK.

[3] Hall, J. W., Masurier, L. E., Baker-Langman, E. J., Davis, J. P. and Taylor, C. A. (2004). A decision-support methodology for performance-based asset management. *Civ. Eng. Environ. Syst.*, 21, No. 1, 51–75.

Chapter 10

Infrastructure Resilience for High-Impact Low-Chance Risks*

Abstract

Infrastructure resilience is the ability of an infrastructure system to withstand or recover quickly from difficult conditions, which in turn requires a detailed understanding of vulnerability and risk. But while designing for foreseeable risks is a challenge, accounting for risks which are difficult or even impossible to foresee — such as those arising from complex interdependent processes — poses a far greater challenge. This chapter argues that civil engineers need a way of addressing such low-chance but potentially high-impact risks if they are to deliver truly resilient infrastructure systems. They need to cultivate a wisdom to admit what they genuinely do not know, and to develop processes to manage emerging unforeseeable consequences. A generalized vulnerability theory that can be applied to any infrastructure system is described, together with an example of how it can be applied to an urban transport network.

Introduction

Physicists define 'resilience' as the ability of an elastic material to absorb energy. Ecologists define it as the ability of an ecosystem to return to its

* This chapter was originally published in *Proc. Inst. Civ. Eng. Civ. Eng.*, 2012, 165, No. 6, 13–19 (co-authors: J. Agarwal and P. S. Godfrey).

original state after being disturbed, while medics refer to an ability to recover readily from illness, stress, depression or adversity. Thus, a general definition of resilience is an ability to withstand or recover quickly from difficult conditions [1], or to adjust easily to misfortune or change.

In the UK government's critical infrastructure resilience programme [2], resilience is defined as, 'the ability of a system or organization to withstand and recover from adversity'. ISO Guide 73 [3] defines resilience as the adaptive capacity of an organization in a complex and changing environment.

According to the UK government, a resilient system or organization will be able to achieve its core objectives in the face of adversity through a combination of good design, protection, effective emergency response, business continuity planning and recovery arrangements [2]. The omission of the word 'quickly' is unfortunate, as speed of response may sometimes be critical.

Resilient Infrastructure

Civil engineers intuitively want to create resilient infrastructure, but until recently few have attempted to articulate what resilience entails. Debates have largely been expressed in terms of 'optimization', 'sustainability', 'robustness', 'vulnerability', 'risk', 'disaster planning' and, more recently, 'complexity' [4]. It is important therefore to clarify similarities and differences between these terms.

At its simplest level optimization is about getting the best out of a system and sustainability is a capacity to endure. A solution that is optimal or highly tuned in one context may well be vulnerable in another.

Vulnerability is a key term which, although referred to a great deal in the literature, is often defined in an unhelpful way that confuses it with risk. In this chapter it is defined, based on previous research (e.g. [5]), as susceptibility to damage or perturbation — especially where small damage or perturbation leads to disproportionate consequences. This is more revealing and helpful in practical usage and distinguishes it from risk much more clearly.

Whereas the terms optimization, vulnerability and robustness are normally used to refer to whole structures, it is also important to recognize them at localized levels and in existing standard design procedures. So, for example, an I-beam optimized only for simple bending becomes vulnerable to lateral torsional buckling. A generic sense of design for resilience requires work at all levels of detail.

Sustainability logically implies resilience. In logic, the inference that A logically implies B is the case except when A is true and B is false. This means that logically it can be inferred that a system is not sustainable when it is not resilient. But if it is resilient, it may or may not be sustainable because there are other factors, such as environmental management and consumption of resources that are needed for sustainability.

In other words, resilience is necessary but not sufficient for sustainability, but sustainability is sufficient for resilience. In logic, a necessary condition is one that is required — a 'must have' — and a sufficient condition is one that is adequate on its own, that is its existence leads to the occurrence of something. For examples of the importance and direct use of these logical terms in engineering, see Refs. [6–8].

Vulnerability is sufficient for the occurrence of lack of robustness. Robustness is the property of being strong, healthy, hardy and able 'to take a knock'. A robust system is strongly or stoutly built — again with an implied sense of endurance. More generally robustness is the ability of a system to persist when subject to changes or perturbations and uncertain conditions.

Resilience must therefore entail or imply robustness and hence robustness is necessary but not sufficient for resilience, since the latter also includes recovery to an original state or to a state which continues to meet an acceptable level of the original purpose of the system. Vulnerability is especially critical in dealing with high-impact, low-chance risks. A system is not robust if it is vulnerable. Optimizing a system without proper attention to robustness can lead to vulnerabilities through unrecognized and hence unexpected modes of behaviour.

In the simple example used above, when designers of I-beams routinely consider all known limit states — including simple bending, lateral torsional buckling and deflections — they are intuitively creating greater robustness. Optimization, resilience, robustness and vulnerability must be clearly distinguished from risk — risk is in the future — it is the chance of an event that may cause harm and the consequences that follow.

The UK Health and Safety Executive (HSE) [9, 10] recognizes there is no such thing as zero risk — no matter how remote a risk might be, it could just turn up. Risks have to be managed to a tolerable level. This means they should not be regarded as negligible or something that might be ignored, but rather as something that needs to be kept under review and reduced still further if possible.

The British Tunnelling Society [11], however, seems to conflate risk and robustness by defining the fundamental objective of the design process as that of achieving a robust design. It continues by stating that a robust design is one where the risk of failure or damage to the tunnel works or to a third party from all reasonably foreseeable causes, including health and safety considerations, is extremely remote during the construction and design life of the tunnel works. However, it does say that high-consequence but low-frequency events that could affect the works or a third party shall also be considered.

In summary, vulnerability entails a lack of robustness. Robustness is necessary but not sufficient for resilience, and resilience in turn is necessary but not sufficient for sustainability. The logic is illustrated in Figure 10.1,

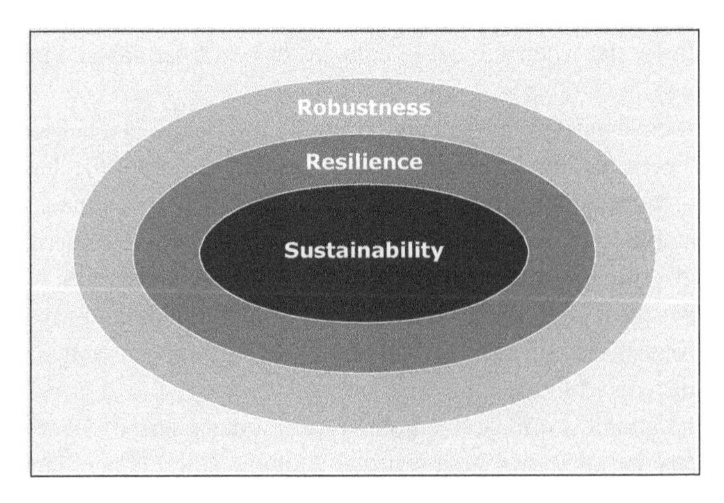

Figure 10.1. Sustainability implies resilience implies robustness.

which shows that the state of being sustainable and not being robust is not allowed. Similarly, the state of being sustainable and not being resilient is not allowed, as is being resilient and not robust. All are handled systemically by managing risks, a well-known example of which is the observational method in geotechnics [12].

As a contribution to civil engineers' attempts to secure resilience in modern complex infrastructure systems, the purposes of this chapter are to

- argue that in complex systems, civil engineers need to cultivate the wisdom to admit to knowing what they genuinely do not know;
- show that civil engineers need consciously to design processes with sufficient resilience to manage unexpected consequences;
- outline briefly a generalization of structural vulnerability theory which can be applied to any infrastructure system.

Complex Systems

Complex behaviour [4] can emerge from interactions between many simpler, highly interconnected processes. There is a growing recognition of new risks through interdependencies that may not be fully understood.

For example, it is known that some (but not all) physical processes are chaotic in the sense that, while they may appear to be reasonably simple, they are inherently difficult to predict. It has been discovered that they may be very sensitive to very small differences in initial conditions and may contain points of instability where paths diverge. This is seen even in quite simple systems, like a double pendulum, as well as bigger and more complex systems like weather forecasting. This is a new kind of uncertainty that presents a new kind of risk.

Highly interconnected systems, such as electrical power supply networks, the Internet, traffic highways and building structures can become vulnerable to quite small damage, cascading to disproportionately large consequences that extend beyond the boundaries of envisaged systems. Even if the chance of the initial damage is very low, the consequences can be very severe. Such systems lack resilience or robustness.

Civil engineers have to recognize that they cannot predict the total behaviour of a complex infrastructure system from the performance of its interdependent parts — they have to expect the unexpected with unintended consequences. An ongoing example is currently occurring in Christchurch, New Zealand where some of the repairs from the earthquake in early 2011 are on hold because people cannot get insurance for new building, unless they are shifting from a damaged house where they were existing customers for an insurance company. Contractors are finding it difficult to insure buildings as they are building them; this is an unexpected and unintended outcome.[a]

Resilience Engineering

Resilience engineering is a term proposed by Hollnagel *et al.* [13] to capture a way of thinking about safety which enhances the ability at all levels of organizations to be robust yet flexible, and to use resources proactively to manage processes to success. They argue that too many people have regarded safety as something a system has (a property) rather than something a system does (a performance).

Resilience engineering therefore abandons the search for safety as a property — such as adherence to standards, calculations of reliability, event trees and counts of human error — and instead sees resilience as a form of control. In other words, the system properties are necessary but not sufficient for safety.

The approach is entirely consistent with methods based on systems thinking proposed by Blockley and Godfrey [6] as a way of 'rethinking construction' after the Egan Report, and with work on human and social factors in man-made disasters by Turner and Pidgeon [14]. By this thinking resilience is an outcome of a process that emerges from the interactions between its sub-processes. So, just as linear elastic strain energy is an outcome, expressed as a property of a material that emerges from the interactions between its molecules, so the resilience of an

[a]Elms, D. G., personal communication (2011).

infrastructure project emerges from the interactions between its well-engineered sub-processes.

A central aspect of vulnerability — and hence robustness, resilience and sustainability — of technical and socio-technical systems is how to ensure that 'surprises' are managed, especially those that have high impact but are of low chance or probability. Incompleteness in risks can be divided into those that are knowable and foreseeable — 'known knowns' — and those which are difficult or even impossible to know and foresee — 'known unknowns' and 'unknown unknowns' [15].

The focus is on clearly identifying, characterizing and managing complex interdependent processes to success by explicitly tracking and managing risk and uncertainty. It is important to stress that these processes are not rigid, inflexible procedures, as implemented in some quality assurance systems. Rather the intention is to create a process model, accessible by an intranet to all involved, which facilitates a rigorous clarity, adaptability and resilience on

- what is being done and on what timescale, particularly for contingency planning?
- who is responsible and how are they accountable?
- what success means and how it can be reached together with a constant monitoring of progress to manage unforeseen unintended outcomes?

Infrastructure systems such as transport, energy, waste, water supply, flood control and information contain complex interdependencies. Experts are naturally and understandably reluctant to consider risks that fall outside the range of their professional scope and expertise. Hence, especially in a commercial context, they can be reluctant to admit when they are operating outside their comfort zone. This can lead to situations where experts reject potential evidence simply because it falls outside their current understanding.

The reason why teams are so important is that they can bring a wider range of skills and expertise than can individuals. But even good teams may be unwilling to give credence to complex and very improbable risks.

If a situation is considered too complicated then there is a danger that organizations may not react at all. Civil engineers need to cultivate the wisdom to admit to what they genuinely do not know [16] so that they can then devise processes that monitor performance with contingency plans to make systems resilient and sustainable, even when subject to unforeseen and unintended demands.

Avoiding Surprises

As stated earlier, one measure of unintended consequences is the number of 'surprises' experienced, particularly surprises that arise from a lack of knowledge or the inability to perceive the consequences of what is known.

In any situation where civil engineers may admit they genuinely do not know something, then they must have robust methods for managing that situation. So, for example, they will need to create processes that develop and consider possible scenarios that have potentially serious consequences, even if they are very unlikely to happen.

Processes are needed to consider how to ensure systems are not brittle, and degrade in a way that, at the very least, allows some control of the safety of people. Processes that build awareness among users of infrastructure systems and in particular for contingency and disaster planning are also needed.

In short, a resilient organization should expect unintended emergent behaviour for novel complex systems and design the systems accordingly [4]. It is not the purpose of this chapter to identify all of these kinds of processes, since that process is itself considerable, but Figure 10.2 provides an outline overview of the processes for repair and recovery.

Grundy [17] has outlined six useful steps to disaster risk reduction.

- Know the hazards and risks.
- Identify weaknesses.
- Retrofit for resilience against all hazards.
- Plan emergency response procedures.
- Educate the community to understand and implement the procedures.
- Rehearse emergency responses regularly.

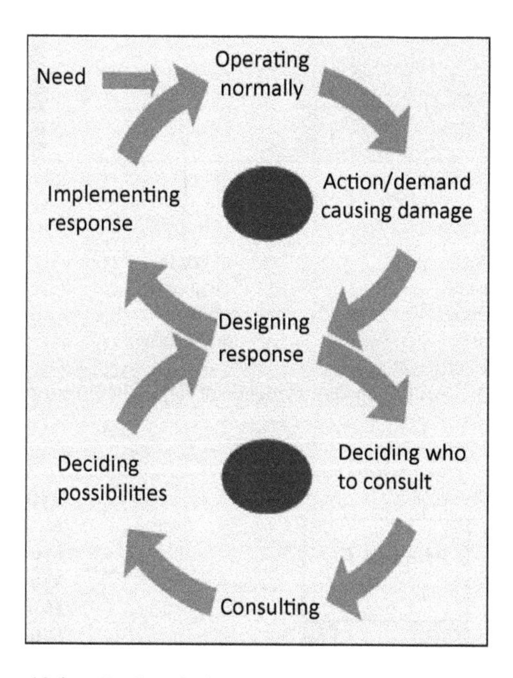

Figure 10.2. Outline design processes for repair and recovery.

These processes are clearly particularly important in seismic zones. For example, retrofitting for resilience must consider checking not only for robust form but also for robust management processes that sense and adapt to respond to threats.

'Italian flags' (Chapter 9, Figure 10.3 and [6]) have been proposed and used to elicit degrees of evidence and to control processes in which evidence is significantly incomplete. Note that the flag is not to be confused with a traffic light — it is a representation of a logical theory of interval probability that includes levels of incompleteness. The green part of the flag indicates the level of positive evidential support for the dependability of a proposition, the red part indicates the level of negative evidential support against the dependability of a proposition and the white part indicates the lack of evidential support for or against the dependability of a proposition, that is the level of incompleteness or 'do not know'.

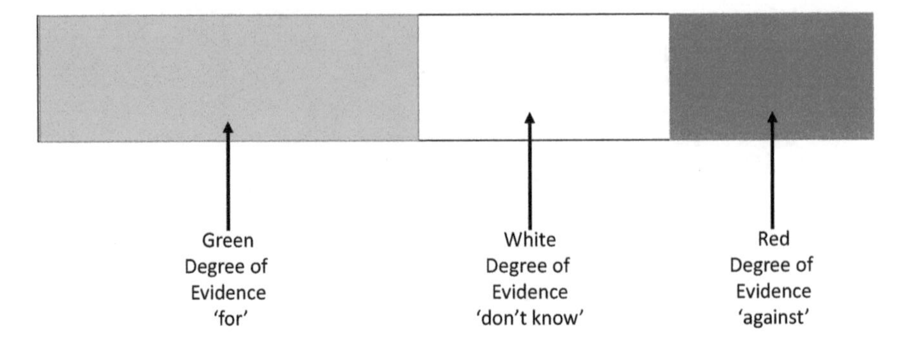

Figure 10.3. An Italian flag evidence of dependability for a purpose.

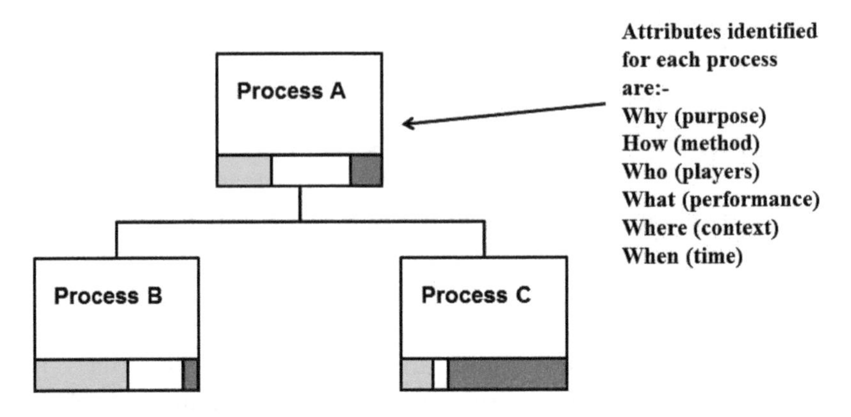

Figure 10.4. Italian flags for some simple processes.

Note: The Italian Flag of evidence for Process C indicates that it is very likely to fail. Unfortunately the eveidence available to the owner of process A indicates that it will be successful. If process C is necessary for Process A then the two process owners need to sort out the reasons for their differing perceptions.

Italian flags can be used entirely informally or formally to support decision-making at various levels in a process hierarchy, as in Figures 10.4 and 10.5, where significant differences of understanding are illustrated. In Figure 10.5, the dam owner has not appreciated the incompleteness of the geotechnical engineer's interpretation of the evidence available or the real worries of the operator. The flags are one way in which these different perspectives are highlighted so that they may be communicated and addressed.

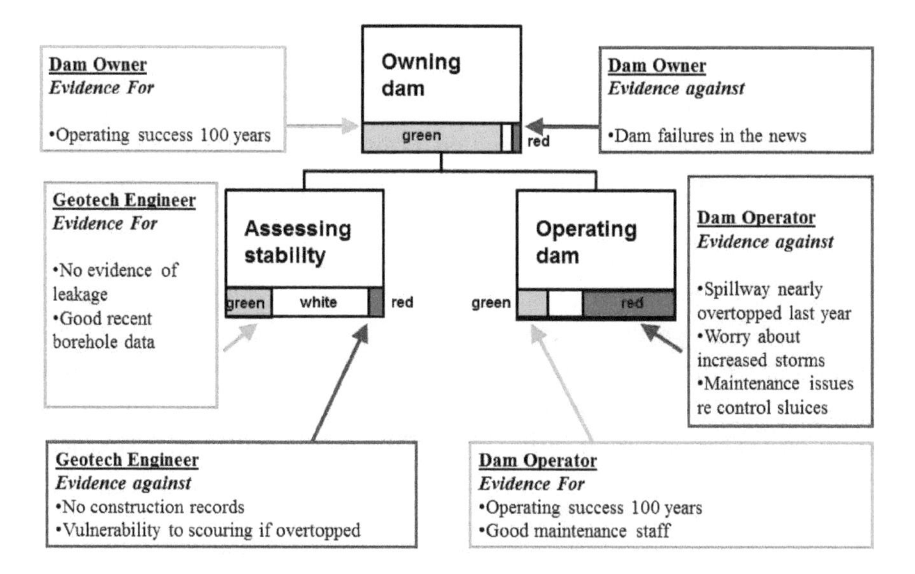

Figure 10.5. Italian flags showing conflicting evidence.

Cascading Failure — A Systems Approach to Vulnerability

Low-chance risks with extremely high consequences are a particularly difficult source of surprises. Eurocode 1, part 1–7 [18], recognizes the need to assess actions arising from accidental human activity, including impact and collisions from wheeled vehicles, ships, derailed trains and helicopters on roofs, as well as gas explosions in buildings.

For nuclear power stations, the UK HSE [9] calls for robustness through redundancy and backup by way of independent components or design diversity, especially in software. It requires a rule of conservatism that pays attention to the quality of a nuclear plant, including management systems, and operational procedures. HSE [10] also calls for the use of the precautionary principle so that where there are threats of serious or irreversible environmental damage, lack of full scientific certainty shall not be used as a reason for postponing cost-effective measures to prevent degradation. This rules out lack of scientific certainty as a reason for not taking preventive action.

Despite such initiatives, the potential for cascading failures from small damage resulting in disproportionate consequences is not well understood. Assumptions of independence that are often made where data are sparse may be seriously misleading. Damage may come from unknown sources and any inherent weaknesses in the form of the system need to be explored at the design stage.

Vulnerability theory [5, 19] is a systems approach to the problem which has been applied to structures and is now being generalized to apply to all engineering systems. There is space here only to present an outline of the theory, in which the form of a system is organized into layers of clusters in a hierarchical process model. The model is then examined for weak points to identify vulnerable scenarios on which risk calculations are based.

Vulnerability may arise because the form of the system has certain characteristics. Form and function are closely related in that an appropriate form is required to achieve a particular function. If the form is damaged then the function will also be affected. Disproportionate consequences derive from a form that is inappropriate because it 'unzips' or cascades when subjected to one or more specific demands, which may not have been anticipated, in an unacceptable way. Hence, vulnerability is examined by concentrating on the way in which the form of a system is affected by any arbitrary damage. Then the results can be combined with the analysis of response to different specific demands.

A system is considered as a set of interacting process objects defined in layers and arranged and connected together in an appropriate form for the purposes of that system [6, 20, 21]. The process objects interact with each other in order to deliver success or to fulfil a role in a higher-level process. They may themselves result from lower-level processes. The nature of objects may differ substantially from one system to another. For example, beams and columns are process objects in a structure and pipes and valves are process objects in a water supply network.

Such systems can be represented as a graph model using nodes and links. The links are the channels of communication between nodes. In most systems there is one channel per link, for example, electrical current or the flow of a fluid such as water. However, there can be more channels along a link, for example, up to six degrees of freedom in a structure. Associated with each link is a parameter describing a quality of the form of the link. This parameter depends upon various components in a system and their relationships, for example, in mechanical and electrical systems [22].

Relationships are expressed in terms of across and through variables. The across-variables (potential) balance around the circuit and the through-variables (flow) balance across any section through the circuit. Table 10.1 summarizes some of the variables for different systems including structures, water supply and traffic networks.

Vulnerability analysis provides a measure of the relative size of the consequences of damage to the effort of producing that damage no matter the chance of it happening — a vulnerability index. The assessment of likelihood of a failure scenario combined with the vulnerability index gives a measure of risk to the form of the system. Clearly this risk may be part of a wider risks assessment and managed within that wider system.

Table 10.1. Parameters governing the form of different systems [22].

Attribute	Electrical circuits	Structures	Traffic	Water pipes	Organizations
Across variable (potential)	Voltage, V	Velocity, v	Need, v	Pressure difference, h	Driver of need and purpose
Through variable (flow)	Current, I	Force, F	Flow, f	Flow, Q	Flow of change
Dissipative component, R	Resistance R $V = R\,I$	Damping c $v = (1/c)\,F$	Resistance to movement	Resistance to flow	Dissipation of energy/ conflict
Across storage component, C (accumulation)	Capacitance C $I = C\,dV/dt$	Mass m (inertia) $F = m\,dv/dt$	Parking	Internal reservoir	Message passing time/inertia
Through storage component, L (delay)	Inductance $V = L\,dI/dt$	Flexibility (inverse of stiffness k) $v = (1/k)\,dF/dt$	Length of link	Length of link	Response time/delay
Weighting parameter describing form of a link, w	Impedance	Stiffness	Transmittance (ease of flow)	(Transmittance (ease of flow)	Impedance

Application to a Transport Network

A road network is vulnerable if a failure in one or more links causes 'knock-on' disproportionate delays to journey times. As indicated earlier, traffic potential — that is the need to travel — drives flows of traffic along links with known transmittance in a directed graph (Figure 10.6). A vulnerability analysis has to relate to delays between chosen reference nodes, for example population centres 2, 22 and 44 in Figure 10.6.

Transmittance depends on capacity speed, length and orientation [23] and is used to calculate a property of a cluster of nodes and links called 'well-formedness' [5].

Well-formedness is a measure of the quality of interconnected loops of nodes within any chosen cluster so that the higher the number of connected nodes with higher transmittance links, the higher the well-formedness.

Well-formedness is effectively an indicator of robust form through the number of good-capacity alternative routes for traffic to flow.

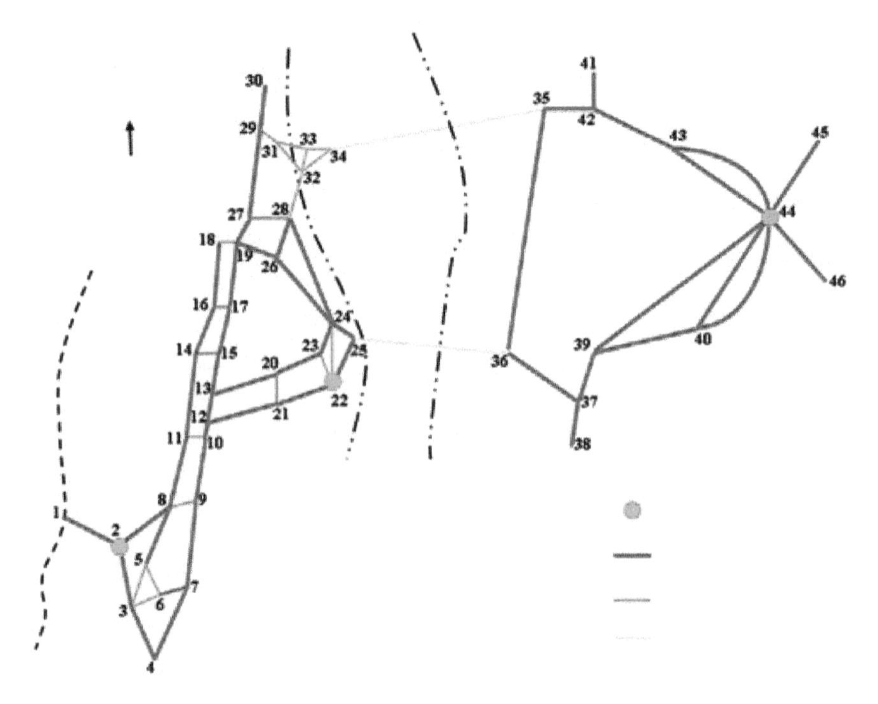

Figure 10.6. An example of a transport network in a coastal city.

For simplicity the network of Figure 10.6 has only three categories of transmittance — A, B and C roads. The hierarchical layers of the systems are created by a clustering process. This starts from the seed loop not linked directly to a reference node and having the highest well-formedness. This seed is grown into a bigger cluster by attaching neighbouring loops if the well- formedness increases. When there is no increase in the well-formedness, a new cluster is seeded.

When all seeds are grown and well-formedness cannot increase further (Figure 10.7(a)) then the clusters are themselves clustered into one

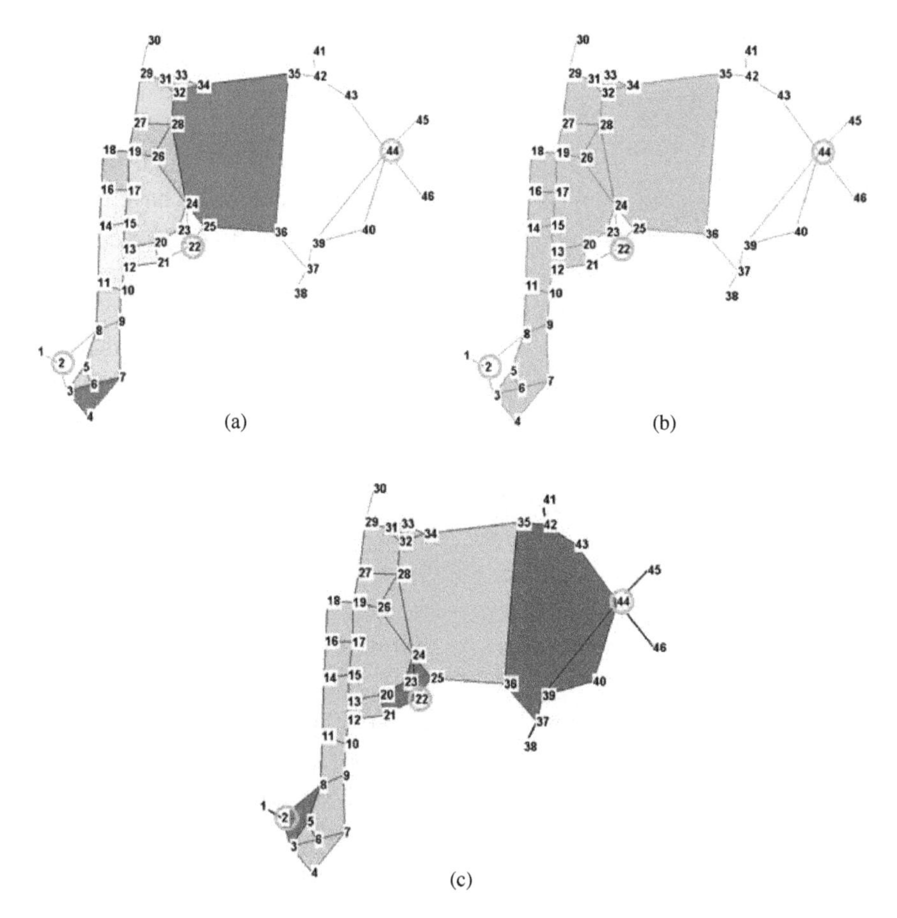

Figure 10.7. A clustering sequence of the results from a vulnerability analysis of transport delays between population centres 2, 22 and 44 in Figure 10.6.

single cluster (Figure 10.7(b)). Finally, the links to the reference clusters are clustered (Figure 10.7(c)). The process produces a natural hierarchy of interconnected clusters. This hierarchy is then systematically searched for various damage scenarios that separate well-formed clusters at all levels in the hierarchy.

For example, and simply for the purposes of illustration, it is straightforward to see in Figure 10.7 that if links 25–36 and 34–35 are cut then reference node 44 is completely separated from the others. Likewise, by cutting links 8–11 and 9–10 then reference node 2 is completely separated. There are other less obvious scenarios which can be prioritized and used when deciding maintenance strategies for the network.

Conclusions

Complex infrastructure systems may contain new risks through interdependencies that may not be fully understood. Civil engineers have to recognize that they cannot predict the total behaviour of a complex system from the performance of its interdependent parts — they have to expect and devise ways of dealing with unexpected and unintended consequences.

Resilience is considered as the ability of a system to withstand or recover quickly from difficult conditions. It is not a simple property like a safety factor or probability of failure; rather it emerges from the interactions between sub-processes. A system is vulnerable if it is susceptible to damage or perturbation, especially where small damage or perturbation leads or cascades to disproportionate consequences.

Vulnerability is sufficient for a lack of robustness. However, robustness is necessary but not sufficient for resilience and resilience in turn is necessary but not sufficient for sustainability. All are handled by systemically managing risks.

Resilience engineering has been used previously to capture a new way of thinking which does not regard safety as something a system has (a property) rather than something a system does (a performance). The approach is entirely consistent with methods based on systems thinking previously proposed by the authors.

One measure of unintended consequences is the number of 'surprises' experienced. Optimizing a particular property of an infrastructure system can increase its vulnerability and reduce its resilience.

Vulnerability theory is a systems approach to this problem which has been applied to structures and is now being generalized to apply to all engineering systems.

References

[1] OUP (1998). *New Oxford Dictionary of English*. OUP, Oxford, UK.

[2] Cabinet Office (2010). *Sector Resilience Plan for Critical Infrastructure*. Cabinet Office, London, UK, Crown copyright, see https://assets.publishing.service.gov.uk/government/uploads/system/uploads/attachment_data/file/271345/sector-resilience-plan-2011.pdf (Accessed September 2019).

[3] ISO (2009). *Risk Management — Vocabulary*. 1st ed. BSI, London, UK, Guide 73:2009.

[4] Elliott, C. and Deasley, P. (eds) (2007). *Creating Systems that Work: Principles of Engineering Systems for 21st Century*. Royal Academy of Engineering, London, UK, Report, see https://www.raeng.org.uk/publications/reports/rae-systems-report (Accessed September 2019).

[5] Agarwal, J., Blockley, D. I. and Woodman, N. J. (2001). Vulnerability of structural systems. *Struct. Saf.*, 25, No. 3, 263–286.

[6] Blockley, D. I. and Godfrey, P. S. (2000). *Doing It Differently*. Thomas Telford, London.

[7] Blockley, D. I. and Henderson, J. R. (1988). Knowledge base for risk and cost benefit analysis of limestone mines in the West Midlands. *Proc. Inst. Civ. Eng., Part 1*, 84, 539–564.

[8] Blockley, D. I. and Robertson, C. I. (1983). An analysis of the characteristics of a good civil engineer. *Proc. Inst. Civ. Eng., Part 2*, 75, 77–94.

[9] Health and Safety Executive (1992). *The Tolerability of Risk from Nuclear Power Stations*. Health and Safety Division, HMSO, London, UK, see http://www.onr.org.uk/documents/tolerability.pdf and http://www.hse.gov.uk/nuclear/ keythemes.htm (Accessed 2 March 2012).

[10] Health and Safety Executive (2001). *Reducing Risks, Protecting People*. Health and Safety Division, HMSO, London, UK, see http://www.hse.gov.uk/risk/theory/r2p2.pdf (Accessed March 2014).

[11] British Tunnelling Society (2003). The Joint Code of Practice for Risk Management of Tunnel Works in the UK. *Institution of Civil Engineers*, London, UK, see http://www.britishtunnelling.org.uk/downloads/jcop.pdf (Accessed 2 March 2012).

[12] Le Masurier, J., Blockley, D. I. and Muir Wood, D. (2006). An observational model for managing risk. *Procs. Insts. Civ. Engs., Civ. Eng.*, 159, 35–40.

[13] Hollnagel, E., Woods, D. and Leveson, N. (2006). *Resilience Engineering*. Ashgate Publishing, Farnham.

[14] Turner, B. A. and Pidgeon, N. F. (1998). *Man-Made Disasters*. 2nd ed. Butterworth- Heinemann, Oxford.

[15] Blockley, D. I. (2009). Uncertainty — Prediction or control? *Int. J. Eng. under Uncertainty: Hazards, Assessment Mitigation*, 1, 1–2.

[16] Government Office for Science (2012). *Blackett Review of High Impact Low Probability Risks*. Government Office for Science, London, UK, Crown copyright, see https://assets.publishing.service.gov.uk/government/uploads/system/uploads/attachment_data/file/278526/12-519-blackett-review-high-impact-low-probability-risks.pdf (Accessed September 2019).

[17] Grundy, P. (2011). *Disaster Risk Reduction: The Engineer's Role*. IEAust, Engineers Australia, Barton, ACT, Australia.

[18] BSI (2008). BS EN 1991-1-7:2008: *Eurocode 1: General Actions — Accidental Actions*. BSI, London, UK.

[19] England, J., Agarwal, J. and Blockley, D. I. (2008). The vulnerability of structures to unforeseen events. *Comput. Struct.* 86, 1042–1051.

[20] Blockley, D. I. (2010). The importance of being process. *J. Civ. Eng. Environ. Syst.*, 27, No. 3, 189–199.

[21] Woodcock, H. and Godfrey, P. S. (2010). *What is Systems Thinking*? International Council on Systems Engineering (INCOSE), Somerset, UK, INCOSE Z Guide 7, see http://www.incoseonline.org.uk/Documents/zGuides/Z7_Systems_Thinking_WEB.pdf (Accessed March 2014).

[22] Shearer, J. L., Murphy, A. T. and Richardson, H. H. (1967). *Introduction to System Dynamics.* Addison-Wesley, Boston.

[23] Liu, M., Agarwal, J. and Blockley, D. I. (2016). Vulnerability analysis of road networks. *Civ. Eng. Environ. Syst.*, 33, No. 2, 147–175. See https://www.tandfonline.com/doi/abs/10.1080/10286608.2016.1148142 (Accessed January 2019).

Part V

Systems Thinking

Preamble

In 1995, I was invited to speak at a conference on risk management held at the Institution of Civil Engineers HQ in Great George Street, London. Patrick Godfrey, who was working at that time for Halcrow (Consulting Engineers), shared the same platform and spoke after me. As I listened to him I realized that here was someone, not an academic but working in industry, who was thinking about risk in engineering in a way very close to mine. We could have been poles apart since he was actively serving his clients as a consulting engineer and I was an 'irrelevant mechanic'. We chatted later over tea and I invited him to Bristol to speak to my undergraduate students.

Over the subsequent years, I got to know Patrick well. He is one of the most complete engineers that I have ever met because he has an ability to communicate across disciplinary boundaries within and without engineering. During his richly varied career as a civil engineer, he has worked in the Seychelles supervising marine works, on the Royal Sovereign Lighthouse, a number of offshore projects in the North Sea, the Gulf, Brazil and New Zealand. He supervised the redesigning and installing of an experimental offshore tower in Christchurch Bay. At the end of the 1980s Patrick, by then Managing Director of Halcrow Offshore, was asked by his CEO to transfer his oil and gas thinking to core civil engineering business. He chose the management of risk as one of his drivers for change. He helped produce the Engineering Council Code of Practice and Guidelines on Risk Issues — a groundbreaking document that emphasizes the human element in risk management. Shortly afterwards he was commissioned, by the Construction Industry Research and Information Association, to produce a client's guide to risk.

Since that time, Patrick has been at the forefront of managing innovative change in the construction industry. He has pioneered new ways of thinking about and managing the complete life cycle of large construction projects. He has developed the interface between business and engineering using a 'soft systems' approach which integrates people with our physical environment. He has worked with clients and designers to understand needs and to find ways of creating projects that satisfy those needs. Patrick believes, as I do, that risk is as fundamental to action as truth is to knowledge. The person who risks nothing, does nothing, has nothing, knows nothing and learns nothing. Only a person who risks is free.

Patrick turned out to be an inspirational role model for the Bristol students — his lectures were extremely popular — the students instinctively knew they were getting material 'straight-from-the-horses-mouth'. I arranged for him to become a Visiting Professor and to take a more formal role by giving some lectures to the fourth year of the Civil Engineering Systems Course. Patrick was unaware, at that time, of systems theory. He soon realized the value of a theoretical basis for some common sense engineering thinking that he was applying to the business of a consulting engineer. As we built our mutual understanding we realized that we should not only pass it on through our teaching but also record it in book form. Thus, by the year 2000 we had developed lectures and turned them into a book which we called *Doing It Differently*. The title came from the very last line of the report chaired by Sir John Egan called 'Rethinking Construction' (1998). The report called for the construction industry to have much more customer focus and to 'do it entirely differently' through committed leadership, integrated processes and teams, a quality driven agenda and a commitment to people. Patrick and I saw systems thinking as the way of delivering the Egan vision.

When I retired from the University in 2006, Patrick left Halcrow to become Professor and Director of the Systems Centre at Bristol University with a thriving Engineering Doctorate stream.

I think Patrick would agree that systems thinking seems to be entirely natural to some and entirely vacuous to others. Opinions seem to polarize into enthusiasts and antagonists. Perhaps we need to appreciate the story of the historical development of systems thinking through reevaluating the relationship between science and engineering as one that moves through the life stages of dependence, independence to interdependence.

I think that the history of engineering naturally divides into five ages — gravity, heat, electromagnetism, information and systems. The first three ages concern our dependence on natural phenomena as well as each other as we harness natural power to improve the human condition. During this time science and engineering developed by leapfrogging over each other as new understanding allowed new tools and new tools allowed more understanding. From the ancient skills used to build pyramids from natural materials to the modern engineering of skyscrapers we have systematically developed our scientific understanding of gravity and used it to build bigger, higher and longer. Our primitive control of fire has developed into mechanical and chemical power from heat through steam, internal combustion and jet engines and manufactured materials. Electromagnetism is a relative latecomer but has given us electricity, motors, computers and telecommunications.

The final two stages have their roots in some kind of cultural disassociation that happened during the Enlightenment. Art and science began to go their separate ways as highlighted by C. P. Snow in 1959 in his 'Two Cultures' Rede lecture. Theory and practice began to fragment. The disassociations are still the topic of much debate. Matthew Crawford in his book *The Case for Working with Your Hands*, argues that each of us is struggling for some measure of self-reliance or individual agency in a world where thinking and doing have been systematically separated. We want to feel that our world is intelligible so we can be responsible for it. We feel alienated by impersonal, obscure forces. Some people respond by growing their own food, some by taking up various forms of manual craftwork. Crawford wants us to reassess what sort of work is worthy of being honoured since productive labour is the foundation of all prosperity. Of course, technical developments will continue but as Crawford argues we need to reassess our relationship with them. Just as a healthy lifestyle is easier if you to have some empathy with your body and how it is performing, so you might drive better if you have some rapport with the workings of your car. In the same way, perhaps a fulfilled life is more likely if you have some harmony with the things you rely on and some feeling of why sometimes they do not perform as you might wish. Of course, just because you rely on something does not mean you must find it interesting. After all few drivers want to know in detail what is happening under the bonnet of a car. However, you can gain a good deal of satisfaction from being able to do something. Practical learning avoids a kind of helplessness so that having to seek assistance becomes a choice rather than a necessity. Skills and talents can help relationships. Practice a talent and you will meet interesting, like-minded people. Whether you are a writer or a real-estate agent, developing abilities can help you in your work. You will be more interesting to others, including prospective clients, and have a greater understanding of the world. Skills help you develop confidence because you are better prepared — you are more resilient in the sense you can act for yourself as you need to. The next **Learning Point No. 16** for Part V therefore is that 'knowing and doing have been systematically separated — we need to bring them together to create resilience'.

As our power to harness natural phenomena has grown so there has been a natural tendency for us to see ourselves as independent of the natural world and later, as we have specialized our knowledge and skills, to become independent of each other. We humans convinced ourselves that we are somehow special and superior to the rest of the natural world which we are free to exploit as we wish.

The age of information did, at first, reinforce the message that knowledge is power. New computing techniques enabled us to exchange and manipulate data and as a consequence we made systems that are much more complex. For example, the computer-based finite element method allowed us to compute forces in structures and other physical systems such as fluids, in ways that were not possible previously. But the age of information has now turned into the age of complex systems. We now realize that the world is more complex than perhaps some appreciated. New risks have arisen that are quite unpredictable. To deal with these new risks we need to learn new lessons. One clear lesson is the need to reverse previous fragmentation and better appreciate our interdependence on each other and on our natural environment. In Part V we will see how 'systems thinking' is helping us to integrate disparate specialisms by seeing tools as physical 'manipulators' of energy embedded in 'soft' people systems.

The complexity arising from the needs of the modern world has required a new coming together of specialists into teams that can tackle large projects in an integrated way. The modern engineer needs to work to achieve long-term sustainable development. Coping with the massively complex interdependencies between systems is one of the major engineering challenges of the 21st century.

One manifestation of complexity is what Conklin and Weil [1] call wicked problems. Schon [2] calls them messy sets of problem issues — the situations that do not seem to yield to easy solutions. You get the feeling they are rather like sorting out a bowl of tangled spaghetti that seems to get more tangled the more you try to sort it out. Conklin and Weil describe a subtle but pervasive pain in organizations. You recognize it, they say, by complaints such as 'I can't get anything done — I've got to go to another meeting', or 'Why does everything take so long these days?' The pain comes from the clash between what we hope, expect and plan to happen and what actually happens — the reality. It comes from not recognizing that some of the problem-solving techniques we have been taught are not adequate for wicked and messy problems. Laurence J. Peter wrote, 'Some problems are so complex that you have to be highly intelligent and well informed just to be undecided about them'.

How familiar do the following statements feel to you? You do not seem to understand the problem until you have developed a solution. There does not seem to be a definitive statement of what the problem is and so there really is no definitive end. There seem to be so many interlocking issues and constraints. There are many unintended consequences of the decisions that are taken. There are so many people involved that the whole thing is a social process which comes to an end when there is some sort of agreement or everybody gets tired. The constraints on the solution change over time, sometimes quite suddenly and rapidly. It is not so much that the goal posts shift — it seems as if you are being required to play a different game.

Change is inevitable and the pace of it seems to increase year by year. You may start a process thinking you are aiming at one set of goals and then, because of changing conditions, you are forced to change direction. It is rather like starting a game of soccer where the aim is to score a goal by kicking the ball under the posts and then being told you are now playing rugby and you should kick the ball over the posts. We are taught throughout our school and higher education that the power of rational thinking is all that we need — that we can solve these wicked problems by linear, logical processes stepping through the stages of the problem in a clear sequence. Of course, this works for many well-formed reductionist (see Chapters 5, 7 and 8) problems, e.g. the analysis and design of idealized structures and other engineering science calculations. Although they probably did not seem simple when we were students struggling to understand, they are simple in the sense that they follow a logical step-by-step sequence to a 'correct' answer. Conklin and Weil [1] call these 'tame' problems because they yield to traditional methods in a reasonable time period. However, if we try to use tame methods on wicked problems, we run into difficulties. It is important, however, to emphasize that we must not 'throw the baby out with the bath water' here! We must not underestimate the importance of linear, logical problem-solving processes. They are *necessary* to develop the mind, they are necessary for a good education and they have enabled us to achieve many magnificent things. If you cannot solve tame simple problems, you will probably not be able to tackle wicked and messy ones. It is just unfortunate that tame simple methods are not *sufficient* for wicked and messy problems. They lead to the organizational pain described earlier. **Learning Point No. 17** is that we must retain the best of the old reductionist approaches as we develop new systems-thinking approaches — in other words, 'integrate reductionism and holism to create synergy'.

I follow this line of argument in Chapter 12 and argue that **Learning Point No. 18** should be that 'delivering resilience will require systems — thinking skills that go beyond technique'. This assertion is founded on Aristotle's notion of *phronesis* as practical wisdom. I argue that practical wisdom has been lost in contemporary thinking and that it needs to be newly recognized, nurtured and promoted. It is the basis for engineering practical rigour or meeting a need by setting clear objectives involving many values (some in 'wicked' conflict) and reaching those objectives in a demonstrably dependable and justifiable way. In Chapter 13, I say that bridges fascinate many people but more than that, in the wider sense, lots of people like to use a bridge metaphor to describe social relationships. So, I use the bridge analogy to explore the link between hard physical and soft social systems. **Learning Point No. 19** is that 'bridges are built by people for people'. Bridge builders must work with and through others — they have to work in teams. Teams work effectively when relationships are good. Relationships connect people — as do bridges — so I call them people bridges.

Finally, in Chapter 13, I say that dealing with the consequences of climate change is such an overwhelming issue that it should create within us a sense of common purpose. We will have to learn to tolerate our differences, use the ethical 'golden rule' whatever our religion and in doing so expect the unexpected. We have to get better at valuing evidence and sorting the 'wheat from the chaff'. The slogan 'Save the Planet' is the most misleading ever. The planet will survive — the question is whether the human race will.

References

[1] Conklin, E. J. and Weil, W. (1999). *Wicked Problems: Naming the Pain in Organisations*. See http://www.accelinnova.com/docs/wickedproblems.pdf (Accessed March 2014).

[2] Schon, D. A. (1983). *The Reflective Practitioner: How Professionals Think in Action*. Temple Smith, London.

Chapter 11

The Age of Systems — Risky Futures*

'We must ensure that this never happens again.' How often we hear these words after an inquiry into a failure. But can they be delivered? With some exceptions in earlier chapters we have concerned ourselves only with engineering achievements. Major disasters are fortunately quite rare but when they do happen they hit the headlines because of large-scale damage and number of people killed.

In this chapter we will contemplate the inevitable gaps between what we know, what we do and why things go wrong. As we explore modern ideas of systems complexity we will see that the gaps are filled by risk. Risk is at the heart of major engineering questions. How do we know what is safe? How safe is safe enough? How do we ensure that the London Eye would not fall over? What are the chances of another Chernobyl or Fukushima? Are engineering failures really failures of engineers? Why do so many major projects seem to be over budget? How do we learn from failure so history does not keep repeating itself?

We know from everyday life that things do not always turn out as we want. This is also true of engineering — indeed making decisions in everyday life has much more in common with engineering practice than may seem at first sight. Both require us to use common sense in solving

---

*This chapter is an abridged version of Chapter 6 in *Engineering: A Very Short Introduction*, 2012, Oxford University Press, Oxford.

problems. We have to decide what we want, what we think we know, how we may achieve what we want, what actually to do and finally what we think might be the consequences. We know if things do not work out as we hoped then we will be affected in all sorts of ways — varying from minor upset to deep and major harm. Engineering decisions however affect many more people than do our everyday ones and may expose them to all kinds of risks — including death. So, quite rightly, what engineers do is closely scrutinized and can, ultimately, be tested in a court of law as a duty of care.

Engineering is practical — it is about creating tools that work properly which simply means that they are fit for their intended purpose. Just as our everyday decisions are constrained by financial, social, political and cultural situations, so are those made by engineers. Perhaps, the most obvious constraint is finance because almost all engineering is a business activity. Engineering activity must be affordable but the predictability of final cost is almost as important as the amount. Politics is an example of a less obvious constraint. For example, natural hazards such as earthquakes and wind storms happen all over the world — but when they occur in regions where buildings have not been properly engineered the consequences are much more devastating than they should be. The large-scale loss of life that happened in Haiti in 2010 could have been avoided if the right technology had been in place — but the reason it was not is a political issue.

Failure is a very human condition — none of us like it but most of us have to deal with it. It is a misunderstanding if one thinks that successful people have never failed — what makes them successful is the way they cope with failure, learn from it and try never to repeat it. There is also a rather disturbing cloud around the mystery, complexity and opacity of modern engineering tools. Part of that cloud is a sense of helplessness when our computers sometimes just decide to behave in ways we do not want or central heating gas boilers break down just when we need them during a cold snap. We have an uncomfortable concern that whilst the power of technology is bringing benefits — it is also bringing a sense of alienation and a feeling of vulnerability with more potential for doing harm.

Of course, we cannot undo the past — but we can do things differently in the future and our understanding of and attitude towards risk has a key part to play. We have to accept that science, technology, engineering and mathematics cannot deliver certainty nor guarantee a future with no failures. But what can be delivered is an acceptable level of risk which is comparable with those from natural events. The problem is that the definition of what is an acceptable risk is not straightforward. We all know that people are killed in car accidents but that does not stop us from driving — most of us consider the balance between risk and benefit to be acceptable. Our perceptions of risk vary greatly in complicated ways. One of the main deciding factors is the degree to which the risk is familiar and we feel in control.

The simple reality is that what we think we know is always incomplete — there is always a gap between what we think we know and what we actually do know — so when we act there is bound to be a gap between what we do and what may happen. Incompleteness (that which we do not know) was famously ridiculed in the media when Donald Rumsfeld, the one-time US Secretary of State for Defense said, 'There are known knowns — these are things we know we know. There are known unknowns … these are the things we know we do not know. But there are also unknown unknowns; these are the things we don't know we don't know'. But Rumsfeld was right. For example, the actual mechanism of failure of the Dee Bridge in 1847 was only understood some 50 years after the event.

How do engineers actually deal with risks? They do their utmost to make it acceptably small in two distinct ways. The first is to make sure the physical tool works properly with a good safety margin. The second is to make sure that people and organizations are well managed and properly controlled so that the risk of error is controlled. Traditionally, these two ways of working are seen as quite distinct — one 'hard' and objective and the other 'soft' and subjective. Engineers know that no matter what they calculate equipment does fail, humans make mistakes and natural hazards, such as earthquakes do occur. So, they often have backup, or contingency plans — this is known as defence-in-depth. The basic idea is to try to prevent an accident in the first place with appropriate safety factors but then to limit the progress and consequences if one should occur.

The engineering team looks at all of the possible demands they can think of and they try to make the chance of failure acceptably small. For example, they want to make sure your car will always start when you want it to — so they look at the reasons why your car might not start. In a similar way engineers examine the safety of a nuclear reactor by drawing enormous logic diagrams covering many pages which trace how an event (such as a pump that fails to circulate cooling water in the nuclear reactor) might affect other parts of the system. These are called event trees. They also draw diagrams that show how a fault may have been caused by other credible faults — these are called fault trees. A partial fault tree for your car not starting is shown in Figure 11.1. Clearly all possible faults are not equally likely so engineers will assess the relative frequencies of faults.

But some tools have become so very complex that it is now impossible to draw an event tree for all possibilities. Engineers have therefore started to work and manage the risks in layers.

The story of earlier chapters tells how the growth and success of engineering and technology has largely been due to our increased understanding of physical phenomena. But what has not advanced so quickly is our understanding of ourselves — how we organize to achieve the things we want. The IT revolution of the 20th century, together with advances in biochemistry and our understanding of the chemistry of DNA are perhaps

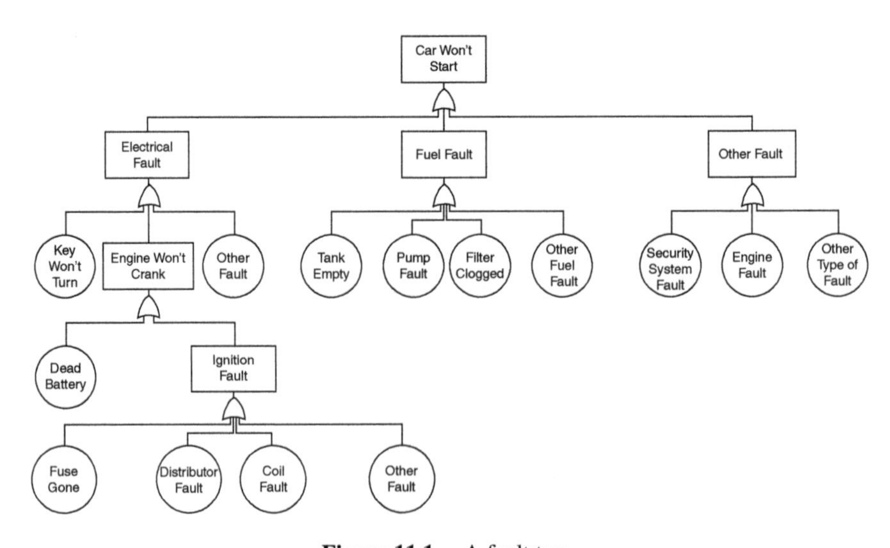

Figure 11.1. A fault tree.

the ultimate expression of the success of reductionist science. But now, in the 21st century, we are beginning to understand how complex behaviour can emerge from interactions between many simpler highly interconnected processes. We are entering into the age of systems with a potential for new risks through interdependencies we may not fully understand.

As a result, some engineers have begun to think differently — they use what many call systems thinking. A system is a combination of things that form a whole. As earlier (Chapter 8) we can distinguish a 'hard system' of physical, material objects, such as a bridge or a computer from a 'soft system' involving people. Hard systems are the subject of traditional physical science. They comprise objects as tools that all have a life cycle — they are conceived, designed, made, used and eventually discarded, destroyed or recycled. They have external work done on them and they respond by doing internal work as they perform. In effect they are 'manipulators' of energy — processes of change in which energy is stored, changed and dissipated in specific ways depending how the elements of the systems are connected. A complex hard system has a layered structure just as in a computer. At any given level there is a layer underneath that is an interconnected set of sub-systems, each of which is a hard system and also a physical process. This kind of hard system thinking enables us to see commonalities between different specialisms that were previously thought to be different. Each object in each layer is a process driven by a difference of potential — an 'effort' to cause a flow which is opposed by impedance. For example, the difference in height between two ends of a water pipe causes the water to flow from the high to low end. The difference in the change of velocity of a mass causes the flow of internal forces. A difference in temperature between two points in a body causes heat to flow from the hot to the cold. A difference in the voltage across two electrical terminals causes a flow of electrical current measured in amps.

Processes interact with each other — we can think of them as sending 'messages' about their own 'state of affairs' to their 'friends and neighbour' processes. The power of a process is the input effort times the resulting flow, e.g. watts are volts times amps. Power is also the rate at which energy is used. In other words, energy is the capacity for work in a process and is an accumulation of power over time. During these processes some of the impedance dissipates energy (resistance), some of it stores potential (capacitance) some of it stores flow (induction).

A crucial question at the heart of all soft systems is 'How do we judge the quality of information on which we depend to make decisions that could risk someone dying? Like the rest of us, engineers want information that they know to be true. If information is true then we can use it without concern. But what is truth? Philosophers have been discussing this since Plato. In engineering, as in everyday life, we need a practical common-sense view that helps us to manage acceptable risks. So, we accept that a true statement is one that 'corresponds with the facts'. But what are facts? Facts are self-evident obvious truths. We have an infinite regress since facts are true statements — we have defined something in terms of itself. In everyday life, for most of the time, this just does not matter. In engineering, because of the duty of care for peoples' lives, we must examine the notion of truth and risk a bit further.

In pre-modern society there were broadly two ways of arriving at truth, mythos and logos. Mythos derived from storytelling. It was often mystical, religious, emotional and rooted in the subconscious mind. It required faith — belief that cannot be proved to the satisfaction of everyone else — and lacked rational proof. Logos, on the other hand, was rational and pragmatic and was about facts and external realities — the kind of reasoning we use to get something done. Church building and church art was a physical expression of faith — the expression of an emotional truth of mythos through the rational tools of logos. To our modern minds, engineering and science seems to spring only from practice based on logos, but the distinction between logos and mythos has never been, and perhaps never will be, one of total clarity. Spiritual faith springing from mythos is genuine and real to the believer and an important basis for the way we live our lives because it provides answers to deep questions about the meaning and purpose of life — but it is not testable in an objective way. Knowledge in STEM is shared objective information that is outside of any one individual. Unfortunately, there are two big problems. First, it has become highly specialized and quite fragmented so that the newest details are understood only by a relatively small number of specialists. Second, it is never simply factual and total but is always incomplete as we discussed earlier through the remarks of Donald Rumsfeld.

Whatever the source of our beliefs, spiritual or practical, mythos or logos, what we do is based on what we believe to be true — in other words on what we think we know. The incompleteness gaps between what we know and what we do and what might be the consequences — intended or unintended are filled by faith. In mythos, this can lead to major tensions between religious sects. In the logos of STEM, faith is a small but inevitable aspect of risk that engineers seek to minimize but can never eliminate. These gaps are often denied, ignored or misunderstood but are actually of critical importance in risk. The job of the engineer is to make the risks acceptably small. In doing so engineers do not look for truth — that is the purpose of science — rather they look for reliable, dependable information on which to build and test their models of understanding. They are acutely aware of context.

The potential that drives the flow of change in a soft system process stems from a need or a want — the answers to our questions *why*. Peter Senge called it a creative tension between 'where we are now' and 'where we want to be in the future'. The flow of change is captured by asking questions and tracking the answers about *who, what, where* and *when*. For example, *who* questions may be about the effects of changes in personnel occupying key roles. *What* questions concern choices, measurements and monitoring of performance indicators including evidence of potential problems and/or success. *Where* questions are issues of context and the impact of changes in context and *when* questions are about timing. *How* questions are about policies, methodologies and procedures — they are the way the change parameters of *who, what, where* and *when* are transformed. One way of envisaging the relationship between these factors is *why = how (who, what, where, when)*. This is not a mathematical formula but is intended to capture the idea that a process is driven by the potential difference of *why* creating a flow of change in *who, what, where, when* through transformations *how*. We can speculate that impedance in soft processes is made up of factors analogous to hard systems. So, resistance is a loss of energy perhaps due to ambiguity and conflict. Capacitance is an accumulation of our ability to do things or to perform. Inductance is our capacity to adapt and innovate. This way of 'systems thinking' is beginning to provide a common language for hard and soft systems though there is still some way to go to make it totally effective and many engineers have not yet embraced it.

How do we deal with risk in this complexity of multiple layers of emergent characteristics? We take a tip from the medics. Forensic pathology is the science or the study of the origin, nature and course of diseases. We can draw a direct analogy between medics monitoring the symptoms of ill health or disease in people with engineers monitoring the symptoms of poor performance or proneness to failure in an engineering system such as a railway network. Structural health monitoring is one existing example where measurements of technical performance of a complex piece of equipment are made to detect changes that might indicate damage and potential harm before it becomes obvious and dangerous. At present its focus is purely technical with axioms that include the assertion that 'all materials have inherent flaws or defects'. This helps us to see that less than perfect conditions may have existed and gone unrecognized in a hard system for some time — there is potential for damage to grow that perhaps no one has spotted. Under certain circumstances those conditions may worsen and it is the duty of those responsible to spot the changes before they get too serious. We can think of these hidden potential threats as hazards, i.e. circumstances with a potential for doing harm or 'banana skins' on which the system might slip. We attempt to detect these hazards by looking for changes in important measurements of performance. For example, steam railway wheel tappers used to check the integrity of steel wheels by striking them with a hammer — a change in the sound told them that the wheel was cracked. Hazard and operability studies (called Hazop) are widely used in designing chemical engineering processes to identify and manage hazards. The safety of a hard system may also be protected by controlling the functional performance of a process automatically. Engineers design into their hard systems feedback tools that operate on the inputs to make the desired outputs. Watt's centrifugal governor (Chapter 3) to control the speed of a steam engine by changing the input flow of steam was an example. Water supply (see Chapter 1), speed controllers on cars, aircraft landing systems and space craft are amongst the many examples where control engineering is now used.

Just as there are technical hazards in hard systems so there are human and social hazards in soft systems. For a soft system, the axiom noted earlier becomes, 'all soft systems have inherent flaws or defects'. Social scientists, such as Barry Turner, Nick Pidgeon, Charles Perrow and Jim

Reason have studied many failures including those mentioned at the start of this chapter. They have discovered that human factors in failure are not just a matter of individuals making slips, lapses or mistakes but are also the result of organizational and cultural situations which are not easy to identify in advance or at the time. Indeed, they may only become apparent in hindsight.

For example, Jim Reason proposed a 'Swiss Cheese' model. He represents the various barriers that keep a system from failing such as good safe technical design, alarms, automatic shutdowns, checking and monitoring systems as separate pieces of cheese with various holes that are the hazards. The holes are dynamic in the sense that they move around as they are created and destroyed through time. The problems arise if and when the holes in the 'cheese' barriers suddenly 'line-up'. In effect they can then be penetrated by a single rod representing a path to failure — a failure scenario. The hazard holes in the cheese arise for two reasons. First are the active failures — unsafe acts by individuals are an example. Second are latent 'pathogens' already resident in the 'cheesy' system. These pathogens result from various decisions and actions that may not have immediate safety consequences but which may translate into later errors — for example, where undue time pressures result in some 'corners being cut' in order to reach a time deadline on a particular project. The pathogens may lie dormant and unrecognized for many years — they may only be discovered when the cheese holes line up.

Barry Turner argued that failures incubate. I have described his ideas using an analogy with an inflated balloon where the pressure of the air in the balloon represents the 'proneness to failure' of a system. The start of the process is when air is first blown into the balloon — when the first preconditions for the accident are established. The balloon grows in size and so does the 'proneness to failure' as unfortunate events develop and accumulate. If they are noticed then the size of the balloon can be reduced by letting air out — in other words, those responsible remove some of the predisposing events and reduce the proneness to failure. However, if they go unnoticed or are not acted on then the pressure of events builds up until the balloon is very stretched indeed. At this point only a small trigger event, such as a pin or lighted match, is needed to release the energy pent up in the system. The trigger is often identified as the cause of the accident but it is not. The over-stretched balloon represents an accident waiting to happen. In order to prevent failure we need to be able to recognize the preconditions — to recognize the development of the pressure in the balloon. Indeed, if you prick a balloon before you blow it up, it will leak not burst. Everyone involved has a responsibility to look for evidence of the building pressure in the balloon — to spot the accident waiting to happen — and to diagnose the necessary actions to manage the problems away.

The problems we are facing in the 21st century require all of us, including engineers, to think in new ways. The challenge for STEM is to protect the important specialisms that allow us to progress our detailed work whilst at the same time providing a set of integrating ideas that allow us to see the big picture, to see the whole as well as the parts — to be holistic but to keep the benefits of reductionist science. Engineering systems thinkers see the many interacting cycles or spirals of change that we have identified in earlier chapters, as evolutionary developments in knowledge (science) and action (engineering) that leapfrog over each other. But it is not the Darwinian evolution of gradual accumulation — it is purposeful human imagination used to improve our quality of life so that to act you need to know and to know you need to act. This view is in direct contrast to a reductionist philosophy that sees knowledge as more fundamental than action. Systems thinkers value knowing and doing equally. They value holism and reductionism equally. They integrate them through systems thinking to attempt to get *synergy* where a combined effect is greater than the sum of the separate effects. It is a new philosophy for engineering. It may sound a little pretentious to call it a philosophy — but it does concern the very nature of truth and action since risk is as central an idea to systems thinking just as truth is to knowledge. Put at its simplest truth is to knowledge as the inverse of risk is to action. The intention of knowledge is to achieve understanding whereas the intention of action is to achieve an outcome. Truth is the correspondence of understanding with 'facts'. Risk is a lack of correspondence of outcome with intended consequences. So, a degree of truth between true and false is analogous to a degree of risk between failure and success.

At the start of this chapter we asked whether we can ensure that a specific failure will never happen again. How do we know what is safe?

How safe is safe enough?' The answer is that we cannot eliminate risk but we can learn lessons and we can do better to make sure the risks are acceptable. We can punish negligence and wrong doing. But ultimately 'safe enough' is what we tolerate — and what we tolerate may be inconsistent because our perception of risk is not straightforward. We can ensure that structures like the London Eye would not fall over, or the risks in a railway signalling systems are acceptable, by making sure that those in charge are properly qualified and really understand what they are doing. But the chance of the London Eye falling over is not zero, the chance of a train crash due to a signalling failure is not zero. The truths of science are not enough to ensure a safe engineered future for us all — engineers use science but there is a lot more to it than simply applying it. On top of that all of the activity is embedded in societal processes in which everyone has a part. In a democracy good choice requires good understanding. Informed debate should enable us to find more consistent ways of managing engineering risk.

Climate change may well be the defining test. The debate has to move on from the questions of whether change is man-made. We don't know for certain — but the evidence is overwhelmingly strong. The stakes are so high that we need to organize ourselves for major weather events which, if we are lucky will not occur. Engineers have to deliver sustainable systems with low throughput of material and energy with more recycling. More attention needs to be given to making systems durable, repairable, adaptable, robust and resilient.

The tensions are pulling the traditional divisions between engineering disciplines in a number of opposite directions and they are creaking under the strain. Specialization has led to fragmentation and a loss of overview. Engineers who hunker down in their silos cannot contribute to challenges that do not fit into traditional boxes. There is still a need for highly specialist expertise that has to keep up with new technology. But perhaps even more importantly, that specialism must be tempered with a much wider understanding of the 'big picture' than has typically been the case in the past.

We will end our story where we began — engineering is about using tools to do work to fulfil a purpose. Over centuries we have created some very large complex interconnected systems that are presenting new vulnerabilities, risks and challenges. Climate change is forcing us to focus on energy — the capacity to do work. The laws of thermodynamics tell us we can shift energy around but we cannot destroy it. However, as we shift it around some of it is lost to us — irretrievably no longer available to do work — we cannot get something for nothing, entropy inexorably increases. The energy performance of buildings is a good example of the need to do better. The UK Royal Academy of Engineering says that too often no one holds an overview and so the engineering solutions sometimes lack coherence. 'Embodied energy' is presently rarely considered — this is the energy used to make all of the materials and components to be used in a particular building before they reach a construction site. Exergy analysis is hardly ever used. Exergy is related to entropy and is a measure of the available work in a system that is not in equilibrium with its surroundings. Using it we can capture both the quality of available energy as well as the quantity. For example, it tells us that it is inefficient to use high-grade electricity from the national grid for low-grade domestic heating that takes us from ambient to around 20°C.

Energy, entropy and exergy are examples of how the challenges of the 21st century require the engineering disciplines that are much better at integrating their expertise to find synergy. If we are to make and maintain highly reliable and sustainable complex systems then we need more of our specialist engineers to be systems thinkers that can deal both with the detail and the 'big picture' — a synergy from the integration of reductionism and holism.

References

[1] Agarwal, J., Blockley, D. I. and Woodman, N. J. (2001). Vulnerability of structural systems. *Struct. Saf.*, 25, No. 3, 263–286.

[2] Blockley, D. I. (2009). Uncertainty — Prediction or control? *Int. J. Eng. under Uncertainty: Hazards, Assessment Mitigation*, 1, No. 1–2, 73–80.

Chapter 12

Finding Resilience through Practical Wisdom*

Abstract

Delivering resilience will require systems-thinking skills that go beyond technique. Relationships between resilience and modern criteria such as risk, vulnerability, robustness and sustainability are outlined. To address them the dominant paradigm of technical rationality needs to be set within the context of reflective practice. Aristotle's notion of *phronesis* as practical wisdom has been lost in contemporary thinking and needs to be recognized, nurtured and promoted. Engineering practical rigour is meeting a need by setting clear objectives involving many values (some in 'wicked' conflict) and reaching those objectives in a demonstrably dependable and justifiable way. Seven elements in practical rigour are described. The key to delivering resilience is to allow professionals to publicly admit uncertainty and in particular to say that we do not know when we *genuinely* and collectively do not know. This will enable us to manage risks by integrating people, purpose and process to improve performance by collaborating and learning together.

*This chapter was originally published in Blockley, D. I., *J. Civ. Eng. Syst., Special Edition on Resilience*, 2014.

Introduction

Infrastructure resilience is the ability of an infrastructure system to withstand or recover quickly from difficult conditions [1]. Those conditions are usually some form of damage with immediate and longer-term consequences that vary from slight to catastrophic. Designing, building and maintaining resilient systems to cope with foreseeable risks is a challenge. Accounting for risks which are difficult or even impossible to foresee — such as those arising from low chance but high consequence risks and from complex interdependent processes with 'wicked' uncertainties (i.e. problems that are difficult or impossible to solve because of incomplete, contradictory, and changing requirements that are often difficult to recognize) pose an even greater challenge.

In this chapter, my purpose is three-fold. First, I want to go beyond technique to address the kinds of systems-thinking skills I think we need to encourage to deliver lifelong resilience. Second, because we face a shortage of recruits into engineering to deal with resilience [2], I want to present some ideas to counter the commonly held view (and not just in the wider non-technical public) that engineering is inherently not creative partly because the word 'technical' implies mere technique, 'system' or 'systematic' implies mere rule following and engineering implies the mere 'application of what we know', i.e. 'applied science'. Thirdly, I want to set out some of the principles that underpin 'reflective practice' (RP) and its relationship with a kind of practical wisdom that transcends 'technical rationality' (TR) (Chapter 5, [3–5]). I will argue that in order to deal with wicked uncertainties we will need to explicitly recognize and promote the practical wisdom and confidence to admit what we genuinely do not know (for example, about the future consequences of climate change) so that we can learn together to evolve processes to manage emerging unforeseen and unintended consequences.

The structure of the chapter follows from these three purposes. Since there is, as yet, no consensus on the nature of resilience I will begin by setting out its relationship with other similar design criteria such as risk, vulnerability, sustainability and robustness. Secondly, I will review the limited literature on practical wisdom and practical intelligence. Thirdly, I will outline the systems concepts needed for thinking about resilience through practical wisdom and I will set out the essential ingredients for the inculcating, encouraging, developing, learning and using of practical wisdom to engineer resilience through the entire life cycle of an infrastructure project.

Resilience

Martin-Breen and Anderies [6] present a useful summary of resilience in which they identify three approaches to providing resilience. The first, they call the engineering approach, is defined as bouncing back and is rather limited. The second, they call the systems approach, is about maintaining functions after disturbance. The third, they call complex adaptive systems approach, is the ability to withstand, recover from and reorganize in response to crises. This third approach, by its very nature, requires a holistic overview which incorporates both hard and soft systems thinking. Dias and Blockley [4] argued that there is a problem that restrains engineers from addressing these wider issues adequately. It is that academic institutions are dominated by TR based on the scientific method (Chapter 5). They urge us to base our thinking on RP as a systems-thinking approach that encourages reflective interaction. RP is wide and holistic (whole is more than the sum of its parts) whereas TR is precise and reductionist (whole is just the sum of its parts). Of course, TR models provide crucial evidential support for RP decision-making but must be interpreted in context.

The challenges of the 21st century require engineers to consider a number of difficult, not easily calculable, criteria such as risk, robustness, vulnerability, sustainability and resilience which go way beyond the partial factor approach within the strength and serviceability criteria of limit state design. In response researchers have developed quantitative scientific models but within a TR, rather than a PR, paradigm. The primary examples are the risk and reliability studies since Freudenthal [7] suggested the idea (see ICOSSAR [8–11, 13–15]). Other examples are the vulnerability of systems to low-chance but high-consequence risks and robustness as the property of being strong, healthy, hardy and able 'to take a knock' [16–18]. The latter is not easily mathematized and has been covered, since 1970 after the Ronan Point collapse in 1968, by codified rules in amendments to the UK Building Regulations and in more recent standards [19] to inhibit progressive collapse.

The TR approach in many of these models results in many of the important and difficult practical considerations being omitted because they are too difficult to include or not seen as relevant. This 'selective inattention' is acceptable as long as it is explicitly recognized that models are context dependent. One consequence is that the models are often inadequate for full implementation in practice. The considerable effort in risk and reliability analysis is a primary example with its partial focus on parameter uncertainty and inadequate recognition and treatment of system (modelling) and human uncertainty [20] and the consequent reluctant uptake within the community of practitioners.

Blockley *et al.* [16] have discussed the relationships between these modern criteria (Chapter 10). For example, resilience must entail or imply robustness and hence robustness is necessary but not sufficient for resilience, since the latter also includes recovery to an original state or to a state which continues to meet an acceptable level of the original purpose of the system. Sustainability logically implies resilience. A resilient system may or may not be sustainable because there are other factors, such as environmental management and consumption of resources, which are needed for sustainability. In other words, resilience is necessary but not sufficient for sustainability, but sustainability is sufficient for resilience.

All of these conditions and criteria are dealt with in practice by steering processes towards success, i.e. by managing risk. In other words, within any given process, hard or soft, decisions are made that will move that process towards success and away from failure. To do this, judgements are made based on evidence from many sources. Sources from the past include project performance, individual experiences and case histories, from the present are current assessments of the 'state of play' of a project, measures of state variables and performance indicators, and finally from predictions of future scenarios using models of all kinds both scientific, heuristic and personal judgements of responsible and appropriately experienced practitioners. The assessment of evidence and the decisions made on the basis of that evidence requires engineering practical competence and expertise.

Elms [21] has argued that engineering education has tended to concentrate on the giving of knowledge rather than an imparting of capability. Capability is something more stable and more endurable than knowledge but tends to be acquired almost incidentally as a by-product of obtaining knowledge. Design courses are perhaps an exception. Elms argues that one of the reasons why capability is not taught is because the very nature of engineering is not well understood. Science is truth oriented whereas engineering is goal oriented. Elms states that wisdom is a quality of the way of looking at things; it is the ability to see the world clearly in a coherent picture. The clarity is simple but not simplistic and depends on strong underlying models. Elms has written, 'A wise person has to have knowledge, ethicalness and appropriate skills to a high degree. There also has to be an appropriate attitude; an ability to cut through the complexity and to see the goal and aims, the fundamental essentials in a problem situation and to have the will and purpose to keep these clearly in focus. It is to do with finding simplicity in complexity. More fundamentally it is to do with world-views and the way in which the person constructs the world in which they operate; which is to say, in engineering, that wisdom is to do with having appropriate conceptual models to fit the situation.'

It is my contention that our western intellectual tradition has lost the idea of practical wisdom. There is a mismatch between the way most people in the general population experience practical life, with all of its uncertainties and risks, and the treatment of practice (and specifically professional practice) by intellectuals and then passed on to the rest of us by various media. Many people (technical and non-technical) value practical skills but are browbeaten into seriously undervaluing them as 'worthy'. The consequences include: experience is downplayed when compared to youthful enthusiasm, theoretical 'academic' knowledge is widely assumed to be superior to 'vocational' practical skills; science predominates engineering, technology is seen as merely 'applied science' and engineering is often identified as a trade with little or no creativity and no connection with its Latin root of *ingeniarius* as ingenious practical problem solving. If we are to recruit a sufficient number of young engineers to tackle the major challenges of the 21st century then we require a change in the 'image' of engineering. In turn that requires a change of culture. But such attitudes are deep seated.

Engineering Practical Wisdom

'Some people who do not possess theoretical knowledge are more effective in action (especially if they are experienced) than others who do possess it.'

Aristotle was writing around 330BC in his *Nicomachean Ethics* [12] and referring to practical wisdom — something that many engineers have quite naturally. Because this idea has been lost, practical wisdom is not just unrecognized, valued and nurtured, but many practitioners are made to feel inferior in comparison with those engineers with more theoretical knowledge. The roots of this loss go way back to the influential thinking of the Greeks but have been reinforced and modified by a number of historical developments. Of course, we should remember that Aristotle lived in very different times and we cannot simply apply his ideas directly — but there are observations that are worth considering in their own right as well as helping us to understand how present-day attitudes have arisen.

Necessarily what follows is a simplified account of complex and profound ideas. Aristotle saw five ways of arriving at the truth — he called them art (*ars, techne*), science (*episteme*), intuition (*nous*), wisdom (*sophia*) and practical wisdom (*phronesis*) — sometimes translated rather misleadingly as prudence. *Ars* or *techne* (from which we get the words art and technical skill, technique and technology) was concerned with production but not action — an important distinction for Aristotle. Art had a productive state, truly (in the sense of correctly) reasoned, with an end (i.e. a product) other than itself (e.g. a building). It was not just a set of activities and skills of craftsman but included what we would now call the fine arts. The Greeks did not distinguish the fine arts as the work of an inspired individual — that came only after the Renaissance. So *techne* as the modern idea of mere technique or rule-following was only a part of what Aristotle was referring to.

Science or *episteme* (from which we get the word epistemology or knowledge) was what Aristotle considered that we know that cannot be otherwise than it is. *Episteme* was of necessity and eternal — it is knowledge that cannot come into being or cease to be — it is demonstrable and teachable and depends on first principles. Later, when combined with Christianity, *episteme* as eternal, universal, context-free knowledge has profoundly influenced western thought. Most modern thinkers would reject the idea of science as being eternal and universal but its legacy lives on in the modern debates between the relative claims of science and religion. As I shall refer to later many engineers and scientists now think of science as sets of context dependent models.

Intuition or *nous* was a state of mind that apprehends these first principles and we could think of it as our modern notion of mind, intelligence or intellect. It is worth noting that one modern interpretation of *nous* is as common sense or practical intelligence. To Aristotle wisdom or *sophia* was the most finished form of knowledge — a combination of *nous* and *episteme*.

Aristotle thought there were two kinds of virtues, the intellectual (such as wisdom, understanding and practical wisdom) and the moral (such as liberality or beneficence and temperance or self-control). He described virtues as dispositions towards feelings as distinct from (a) the feelings themselves like desire, anger or fear and (b) our faculties to be capable of those feelings. Dispositions are attitudes, outlooks or world-views that imply choice — the kind of disposition that derives from the way we behave, from habit (*ethos* — from which we get the word ethics) and the way we deal with situations. He wrote 'like activities produce like dispositions'. Hence, we must give our activities a certain quality, because it is their characteristics that determine the resulting dispositions." A virtue writes Aristotle 'is a purposive disposition, lying in a mean that is relative to us and determined by a rational principle and by that which a prudent (i.e. having practical wisdom) man would use to determine it.' A virtue is the pursuit of *arête* (goodness, excellence, efficiency), i.e. what is right and good. Virtue renders something good and makes it perform well. Aristotle wrote 'For example, the excellence of the eye makes both the eye and its function good (because it is through the excellence of the eye that we see well) … human excellence will be the disposition that makes one a good man and causes him to perform his function well.'

Practical wisdom or *phronesis* was an intellectual virtue of perceiving and understanding in effective ways and acting benevolently and beneficently. However, there seems to be some controversy [22] between Aristotelian scholars about whether *phronesis* is only responsible for what contributes to the goal. Aristotle wrote, 'practical intellect does not tell us what ends to pursue but only how to pursue them; our ends themselves are set by our ethical characters'. However, he makes clear that 'man's function depends upon a combination of prudence and moral virtue; virtue ensures the correctness of the end at which we aim, and prudence that of the means towards it'. Elsewhere he writes, 'prudence (practical wisdom) is a virtue not an art'. It necessarily involves ethics, is not static but always changing, individual but also social and cultural. As an illustration of the quotation at the head of this section, Aristotle even referred to people who thought Anaxagoras and Thales were examples of men with exceptional, marvellous, profound but useless knowledge because their search was not for human goods (in the sense of worth or benefit).

Aristotle thought of human activity in three categories *praxis, poeisis* (from which we get the word poetry) and *theoria* (contemplation — from which we get the word theory). The intellectual faculties required were *phronesis* for *praxis*, *techne* for *poiesis* and *sophia* and *nous* for *theoria.*

It is important to understand that in the minds of the ancient Greeks *theoria* had total priority because *sophia* and *nous* were considered to be universal, necessary and eternal but the others are variable, finite, contingent and hence uncertain and thus inferior. It is also interesting that *praxis* was the word for practice that included culturally shared knowledge of how to do it plus informal rules, techniques and case histories. Nevertheless, it required *phronesis* as we will discuss in a moment. *Poiesis* was the act of making or producing something specified — a transformation that continues the world and, perhaps surprisingly for the modern mind, included poetry. Heidegger [23] the influential 20th century philosopher who controversially argued for the 'primacy of practice' over science was later to describe *poiesis* as 'a bringing-forth' or 'a revealing' — something done for a purpose where the product is not simply an end in itself but only a relative or particular end such as a pair of shoes is an aid to a good life because X needs shoes. *Poiesis* relies on *techne* or instrumental moves that produce predictable results that can be organized into a technology, i.e. a *logos* (rational principle or logic) of *techne* — not something we now typically associate with poetry.

How do we characterize what Aristotle meant when he referred to *phronesis*? Aristotle wrote, 'full performance of man's function depends on a combination of practical wisdom (phronesis) and moral virtue; moral virtue ensures the correctness of the end at which we aim and practical wisdom that of the means towards it'. Moral virtue makes the goal right; *phronesis* is responsible only for what contributes to the goal. That is, practical intellect does not tell us what ends to pursue, but only how to pursue them; our ends themselves are set by our ethical characters. As I interpret it *phronesis* is an intellectual virtue or a competence, an ability to deliberate rightly about what is good in general, about discerning and judging what is true and right but it excludes specific competences (like deliberating about how to build a bridge or how to make a person healthy). It is purposeful, contextual but not rule-following. It is not routine or even well-trained behaviour but rather intentional conduct based on tacit knowledge and experience, using longer time horizons than usual, and considering more aspects, more ways of knowing, more viewpoints, coupled with an ability to generalize beyond narrow subject areas. *Phronesis* is concerned with 'the capacity for determining what is good for both the individual and the community'. A professional *phronimos* continually mediates between the abstract, theoretical universal and the practical and particular, between generalizations supported by cultural understandings and specific responses to the particularities of the issues or events at hand. A *phronimos* is not a bearded sage who knows all but rather is someone who has learned lessons by reflecting on extensive experience of life with all of its risks and uncertainties — including unintended consequences.

Phronesis was not considered a science by Aristotle because it is variable and context dependent. It was not an art because it is about action and generically different from production. As I have said, art is production that aims at an end other than itself. Action is a continuous process of doing well and an end in itself in so far as being well done it contributes to the good life.

So *phronesis* is not knowledge but it is concerned with the gap between theory and practice. Aristotle thought it possible to be knowledgeable but foolish — 'the mere possession of knowledge … does not make us any more capable of putting our knowledge into practice'. Cleverness, he wrote, is necessary but not sufficient for *phronesis* — something that is not separable from person or group. It is a meta-process or 'outside of the box' thinking of expanding one's context by deepening, elevating, broadening, elaborating perceptions, looking to the future, considering one's values and being altruistic. *Phronesis* is self-referential and self-reflective — each act of *phronetic* judgement is concerned with and relates to the one doing the judging — the *phronimos* is responsible and must be capable of reflecting the various ways in which he/she is implicated.

Long [24] argues that an ontology (the philosophy of being or nature of existence) directed by *phronesis* rather than *sophia* (as it currently is) would be ethical, would question normative values, not seek refuge in the eternal but be embedded in the world and be capable of critically considering the historico-ethical-political conditions under which it is deployed — its goal would not be eternal context-free truth but finite context-dependent truth. *Phronesis* is an excellence *(arête)* and capable of determining the ends. The difference between *phronesis* and *techne* echoes that between *sophia* and *episteme.* Just as *sophia* must not just understand things that follow from first principles but also things that must be true so *phronesis* must not just determine itself towards the ends but as *arête* must determine the ends as good. Whereas *sophia* knows the truth through *nous, phronesis* must rely on moral virtues from lived experience.

In the 20th century, quantum mechanics has required us to think about *sophia* differently — to recognize that we cannot escape uncertainty. Whether we think science is eternal, universal, value and context free *sophia* is now a matter of faith (by which I mean belief not based on proof). Some philosophers such as Cartwright [25] and Bailer-Jones [26], [27] have begun to argue what many engineers have always implicitly assumed, i.e. that science is about incomplete modelling for a purpose. Practitioners, whether through everyday life or professionally through medicine and engineering know now that we cannot escape 'wicked' uncertainties. They are 'writ large' for engineers when considering the possible future possibilities if predictions of climate change actually come to pass. There are no guarantees — Sellman [28] writes that a *phronimo* will recognize not knowing our competencies, i.e. not knowing what we know and not knowing our uncompetencies, i.e. not knowing what we do not know. He states that a longing for *phronesis* 'is really a longing for a world in which people honestly and capably strive to act rightly and to avoid harm … and he thinks it is a longing for *praxis*. We want the good for each person to be done in uncertain practical circumstances and we want the good for humankind to be done, even though we cannot guarantee that these goods will be done' — this is what he means by *praxis.*

In summary, we should perhaps reflect on Aristotle's words, 'it might seem paradoxical that practical wisdom (phronesis), though inferior to wisdom (*sophia*) should prove to be more authoritative'. I think that this intellectual inheritance, i.e. the dominance of context-free, eternal and universal 'True' scientific knowledge has now to be openly challenged — we have to help the general population (and the media) to appreciate the kinds of 'wicked' uncertainties that we face in the 21st century — particularly with respect to the effects of climate change and the complexity and vulnerability of interdependencies within modern engineered systems like the Internet or air-traffic control [29]. Science and engineering (together with medicine and all practical activity that uses science) have different aims but they are equals that depend on each other because 'to know we must do and to do we must know' — history tells us that. I think that one way (and perhaps the only way) of dealing with them is through a collective humility in the face of the challenges. We need a new approach based on a *rigorous collaborative* 'learning together' informed by the recognition, appreciation and exercise of practical wisdom.

Engineering Practical Rigour

Engineering method is often criticized by pure scientists as non-rigorous because engineers use approximations and judgement. In fact, engineers must be rigorous for two impelling reasons. Firstly, engineering products will inevitably be subject to the ultimate test — that of mother nature. If a bridge structure is inadequate to take the forces imposed on it then it will collapse. This imposes a kind of 'natural honesty' since it is a requirement that cannot be twisted by propaganda or 'spin'. Secondly, engineers have a legal duty of care to society. Under this duty engineers have to justify their decisions, if called upon to do so (e.g. when something goes wrong), in a society that questions expertise.

Rigour is the strict enforcement of rules to an end. Mathematical logic is the ultimate form of absolute rigour: it has one value — truth. It is top-down reasoning [13] in which theorems are deduced from axioms (including rules) which are true by definition (but of course may or may not correspond to reality — an example is the difference between Euclidean and spherical geometry). Physical science is bottom-up reasoning since it aims at precise truth as a correspondence to the facts. Hypotheses and theories are conjectured and tested against them — but, as we saw earlier, they have to be set in a context.

Practical rigour is much more complex. It is meeting a need by setting clear objectives involving many values (some in 'wicked' conflict) and reaching those objectives in a demonstrably dependable and justifiable way. I see seven elements in practical rigour. They are as follows:

(1) **Making it work:** Creating practical solutions to meet explicit needs and delivering a system valued in a variety of ways, not just cost.
(2) **Creating appropriate models:** Working with nature — making sensible approximations that respect nature. Far from being the cause of loss of rigour, as our academic accusers may hold, the approximations of our models are the sources of the practical rigour required to create a solution that meets the needs. Practical rigour requires diligence and duty of care that leaves no stone unturned with no sloppy or slip-shod thinking.
(3) **Considering the whole as well as the parts:** The scientific approach is one where we look at a problem, break it down into its separate components, take out the difficult bits that we do not know how to solve and focus on what we can solve. It is a process of selective inattention. Practical rigour does not have that luxury — it requires a rigour that deals with the bits of the problem that we do not always understand too well.
(4) **Making judgements:** Professional opinions are not arbitrary, they are based on Karl Poppers objective world 3 evidence [30] of varying dependability. Opinion based on experience may be less dependable than measurement or standard theory but it has to be testable against world 3 objective knowledge, ultimately, perhaps, in the courts.
(5) **Exercising creative foresight:** Practice requires the creativity to imagine what might happen — how physical things will respond and how people might behave in future situations or scenarios.
(6) **Developing and evaluating dependable evidence:** The only clear way to judge the dependability of evidence is to subject it to as many tests as seems appropriate.
(7) **Feedback and learning:** One of the seven habits of successful people identified by Covey [31] is learning to improve or self-renewal.

Practical rigour implies practical intelligence which in turn implies practical experience. In other words, experience is necessary but not sufficient for practical intelligence which in turn is necessary but not sufficient for practical rigour because practical intelligence and rigour require reflective learning and development on that experience. Thanks to the advances in brain science we can now observe practical intelligence at work. For example, we can see how connections between neurons in the brain develop as someone learns to play the piano [32]. This is a clear demonstration of how practical skills change the patterns of connections between brain cells or neurons, i.e. brain power. Winston [33], the TV presenter and academic medical practitioner describes how learning something new means rearranging the way our brains work — making and strengthening new connections in new neural pathways through the approximately100 billion neurons in our brains. Casey *et al.* [34] state that the human brain undergoes significant changes in structure and functional organization across the life span. They conclude that the shift is an experience-driven maturational process that reflects fine-tuning of neural systems with experience and development but that more research is needed on how learning affects these patterns.

How do we distinguish practical wisdom, practical intelligence and practical rigour? First, it is clear that practical intelligence and rigour are necessary for practical wisdom — but they are not sufficient — it requires more. Practical wisdom is not just about following rules no matter how rigorous. For example, it requires identifying and minimizing the risk of the unintended consequences of the decision. Experienced practitioners know that rules take them only so far. Rules cannot say how to interpret information and to balance evidence and conflicts — especially in complex structures.

One recent non-engineering example of how things can go seriously wrong when practical wisdom is lost has been described by Schwartz [35] as part of the reasons for the 2008 banking collapse. Officials of IndyMac bank in the USA did not think they were doing anything wrong when they wrote dubious loans and were selling them on in a hotly competitive market — profiting from risky loans seemed normal. It was profitable but was it right? Is the traditional aim of banking about giving a service or making money — no matter how as long as it is legal? Another non-engineering example is how a medical doctor should balance empathy for a patient with a necessary detachment. How should a practitioner tell the truth and yet be kind? These questions are not a matter of right and wrong but choice amongst competing right things to do or the least bad or the least worse or the least unhappy outcome. There is no 'one size fits all' answer. Knowing purpose only is not enough. The 'road to hell is paved with good intentions'. Translating intention to acts requires expertise in imagining consequences and what is possible. It requires not just technical or artistic skill but also moral skill — a skill that combines will and skill.

Integrating People, Purpose and Process

So far, we have focused on the individual but we all work with and through people in organizations of various kinds. As stated earlier (Chapter 8 and [36]) most people tend to regard the technical (hard systems physical behaviour) and the managerial (soft systems human behaviour) as totally different in nature. Indeed, the latter is often said not to be engineering at all — just management. This is, in my view an enormous category mistake. In the practice of engineering the two are intimately linked and many problems are located at their interface. The more they can be integrated, the larger the benefits that can accrue. It is an illusion, in my view, to think that we can isolate hard systems modelling from the purpose and context in which the decisions we take that are based on that modelling. I argue that all engineered hard systems are embedded in soft systems because it is people who give them purpose — they design, build and use them. Engineered products do not just exist at a point in time; they develop over time as they are conceived, planned, designed, constructed, used and decommissioned — they are processes. So hard and soft systems are not irreconcilable they are intimately related. Efficient and effective integration requires an understanding and a responsibility that embraces both sides of the boundary to achieve collaboration and shared understanding between disciplines. In short, we need to integrate people, purpose and (*old*) process. I distinguish between the *old* and commonly held view that a process is simply a transformation to a new idea of *new-process* as a fundamental concept that we can use to integrate almost all aspects categorized as *why, how, who, what, where* and *when*. My own approach is put at its simplest, systems thinking is joined-up thinking.

It is getting the right information (*what*) to the right people (*who*) at the right time (*when*) for the right purpose (*why*) in the right form (*where*) and in the right way (*how*). The three ideas at the heart of delivering systems thinking are thinking in layers, thinking in connected loops and thinking about *new-processes*. Everything has life cycle and hence is a process — but one that is set in the context of a system containing other connected processes — some at higher and some at lower levels of definition. The three root principles to do this [36, 37] are (a) finding a hierarchy of levels (e.g. using the concept of a holon), (b) modelling interactions through interconnections (e.g. using graph theory) and (c) modelling processes (e.g. using *why* = *how (who, what, where, when)*). Naturally, there are many other ways of expressing those three root principles — but pushed to express them in one sentence 'I look to model interacting process holons'. Figure 12.1 is an attempt to capture how people, (old) process and purpose can be integrated to improve performance as we attempt to deal with 'wicked' uncertainties.

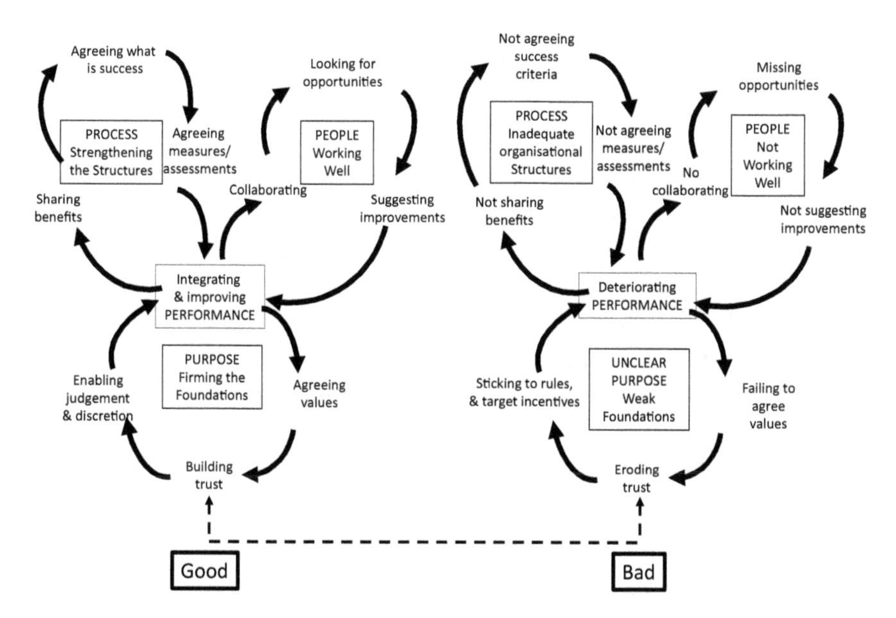

Figure 12.1. Integrating purpose, (old) process and people to improve performance.

Consequences and Conclusions

There are six important consequences of the proposed way of dealing with 'wicked' uncertainties. There is a need for all of us to do the following:

(1) We should be able to publicly admit uncertainty and in particular to say that we do not know when we *genuinely* do not know. Part of the difficulty here has been the lack of scientific consensus on the nature of uncertainty, the understandable and historically determined human need to search for and claim certainty without necessarily recognizing the central role of faith (by that I mean belief not based on proof). Consequently, it is hardly possible for anyone in public or professional life to admit to 'not knowing' because of our relatively recent (historically) collective loss of our trust in 'experts'. All professionals (including politicians) find it near impossible to admit to not knowing as they would stand accused of incompetence and loss of credibility, votes, business clients or patients. Furthermore, they make themselves liable to legal action for incompetence if something does actually go wrong even when there is no proof of negligence under the law of tort.
(2) The emphasis on prediction that is based on the strong performance of TR over the last 100 years or so has to be moderated. For example, civil engineering projects organized as 'predict and provide' [37] cannot deliver success when uncertainties are wicked. That is not to say that scientific prediction is not important rather that it has to be part of the uncertain evidence that we use to manage projects to success.
(3) We should recognize that collaboration rather than confrontation is required. For example, Sir John Armitt (Chairman of the Olympic Delivery Authority for the very successful London 2012) is reported [38] as saying, 'We need to find ministers, who are prepared to say to their departments, "You are free to make mistakes. You are free to mentally allocate some of what you are doing to the 70/30 projects where, in fact, there is a good 30% chance that it will not work — but the 70% is worth going for, so let's go for that. If it goes wrong, you won't be hanged, but you will actually be praised for having a go.

Because we are willing to take a risk, there will be certain things that will be successful'". He states that what was different about the London 2012 Olympics was the level of collaboration across all the projects. This came because of the recognition that to deliver these projects successfully, the client had a responsibility to provide the leadership. If you want to innovate you have to accept that there are risks and so you need flexibility in your budget.

(4) We should recognize that all parties in a wicked problem have to 'learn together'. Godfrey [39] has written, 'As systems engineers we are talking about complex relationships where much complexity is caused by irrational and inconsistent components — mainly people. The complexity faced by engineers is increasing'. Globalization, the need for sustainability, the need to manage customer expectations, the need to recognize interdependence, the opportunity to improve through synergy (by which he means making the whole more than the sum of its parts) rather than aiming for compromise, the need for step change reductions in waste with improvement in sustainability can be achieved through integration of infrastructure processes.

(5) Those responsible for the education and training of engineers should cover the problems of decision-making under uncertainty much more rigorously than has traditionally been the case. In other words, students should be exposed to the ideas of RP, *phronesis* as practical wisdom and practical intelligence and rigour in contrast to logical and mathematical rigour.

(6) Our journey to a level of resilience sufficient to cope with a changing climate requires us to steer a path through a minefield of future hazards. It requires an *evolutionary* observational approach that values 'practical wisdom' and 'learning together' [40]. One way in which engineers have already evolved such a methodology is in geotechnical engineering. Nicholson *et al.* [41] has described the two approaches to design 'predict and provide' which he calls the 'predefined design method' and the Observational Method. The Observational Method originally defined by Peck [42], continuously manages the uncertainty associated with the ground conditions. Observational Method designs are flexible, able to be adapted to suit

the actual conditions encountered during construction. In order to achieve this in a robust way, system of observation and feedback is put in place along with contingency plans, developed in advance, to cater for the most unfavourable conditions. The Observational Method relies on many features of the systems approach; this can be an advantage and a handicap. The advantages of systems thinking in terms of understanding the interconnectedness of the processes of design and construction and establishing a learning process through feedback are clear. A handicap arises when there is a lack of appreciation among the players on a construction project of the need to treat the Observational Method in a systemic way; it may fail if applied partially or in a reductionist way. In short it requires practical wisdom.

References

[1] Cabinet Office (2010). *Sector Resilience Plan for Critical Infrastructure.* Cabinet Office, London. See https://assets.publishing.service.gov.uk/government/uploads/system/uploads/attachment_data/file/271345/sector-resilience-plan-2011.pdf (Accessed September 2019).

[2] Jason Ford (2014). UK 'faces 36,800 shortfall in qualified engineers by 2050', http://www.theengineer.co.uk/channels/skills-and-careers/uk-faces-36800-shortfall-in-qualified-engineers-by-2050/1015304.article (Accessed on June 2019).

[3] Blockley, D. I. (1992). Engineering from reflective practice. *Res. Eng. Des.*, 4, 13–22.

[4] Dias, W. P. S. and Blockley, D. I. (1995). Reflective practice in engineering design. *Proc. Inst. Civ. Engrs. Civ. Eng.*, 108, 160–168.

[5] Schon D., (1983). *The Reflective Practitioner.* Basic Books, New York.

[6] Martin-Breen, P. and Anderies, J. M. (2011). Resilience: A literature review, https://opendocs.ids.ac.uk/opendocs/bitstream/handle/123456789/3692/Bellagio-Rockefeller%20bp.pdf?sequence=1&isAllowed=y (Accessed on September 2019).

[7] Freudenthal, A. M. (1947). The safety of structures. *Trans. ASCE*, 112, 125–180.

[8] IASSAR (2014). *Int. Ass. Struct. Saf. Reliability*, http://www.columbia.edu/ cu/civileng/iassar/ (Accessed on June 2019).

[9] Melchers, R. E. (1999). *Structural Reliability Analysis and Prediction.* Wiley, Hoboken.

[10] Ditlevsen, O. and Madsen, H. O. (2007). *Structural Reliability Methods*. Internet Edition 2.3.7. John Wiley & Sons, Chichester.

[11] Benjamin, J. R. and Cornell, C. A. (1970). *Probability Statistics and Decision for Civil Engineers*. McGraw-Hill, London.

[12] Aristotle (1955). *The Nicomachean Ethics*. Translated by Thomson, J. A. K. Penguin Books, London.

[13] Blockley, D. I. (1980). *The Nature of Structural Design and Safety*. Ellis Horwood, Chichester.

[14] Blockley, D. I. (2008). Managing risks to structures. *Proc. Inst. Civ. Eng., Struct. Build.*, 161, No. SB4, 231–237.

[15] ISO (2009). *Risk Management — Vocabulary* (Guide 73:2009), Ist edn. International Organization for Standardization BSI, London, UK.

[16] Blockley, D. I., Agarwal, J. and Godfrey, P. S. (2012). Infrastructure resilience for high-impact low chance risks. *Inst. Civ. Engrs., Civ. Eng. Special Issue*, 165, No. CE6, 13–19.

[17] England, J., Agarwal, J. and Blockley, D. I. (2008). The vulnerability of structures to unforeseen events. *Comput. Struct.*, 86, 1042–1051.

[18] Government Office for Science (2012). *Blackett Review of High Impact Low Probability Risks*. Government Office for Science, London, UK. See https://assets.publishing.service.gov.uk/government/uploads/system/uploads/attachment_data/file/278526/12-519-blackett-review-high-impact-low-probability-risks.pdf (Accessed September 2019).

[19] British Standards Institution (2006). *BS-EN 1991-1-7, Eurocode 1–Actions on Structures*, Part 1–7. General Actions — Accidental Actions, London.

[20] Blockley, D. I. (Ed.) (1992b). *Engineering Safety*. McGraw-Hill, London.

[21] Elms, D. G. (1989). Wisdom engineering: The methodology of versatility. *Int. J. Appl. Eng. Edu.*, 5, No. 6, 711–717.

[22] Moss, J. (2011). *Virtue Makes the Goal Right: Virtue and Phronesis in Aristotle's Ethics*, https://brill.com/view/journals/phro/56/3/article-p204_2.xml (Accessed on June 2019).

[23] Heidegger, M. (1977). *The Question concerning Technology and Other Essays*. Translated by Lovitt, W. Harper Perennial, New York, pp. 3–35, http://simondon.ocular-witness.com/wp-content/uploads/2008/05/question_concerning_technology.pdf (Accessed on June 2014).

[24] Long, C. P. (2002). The ontological reappropriation of phronesis. *Continental Phil. Rev.*, 35, 35–60.

[25] Cartwright, N. (1999). *The Dappled World: A Study of the Boundaries of Science*. Cambridge University Press, Cambridge.

[26] Bailer-Jones, D. M. (2003). When scientific models represent. *Int. Stud. Phil. Sci.*, 17, No. 1, 59–74.

[27] Bailer-Jones, D. M. (2009). *Scientific Models in Philosophy of Science.* University of Pittsburgh Press, Pittsburgh.

[28] Sellman, D. (2012). Reclaiming competence for professional phronesis. In Kinsella, E. A. and Pitman, A. (Eds), *Phronesis as Professional Knowledge.* Sense Publishers, The Netherlands, pp. 115–130.

[29] Ottino, J. M. and McCormick, R. R. (2004). Engineering complex systems. *Nature*, 427, No. 6973, 3999, http://mixing.chem-biol-eng.northwestern.edu/papers/nature/Nature_Concept.pdf.

[30] Popper, K. R. (1979). *Objective Knowledge.* Oxford University Press, Oxford.

[31] Covey, J. (2005). *The Seven Habits of Highly Effective People.* Simon & Schuster, London.

[32] Blockley, D. I. (2001). Thinking outside of the box. *Struct. Eng.*, 79, No. 20, 22–29.

[33] Winston, R. (2014). How We Learn — Synapses and Neural Pathways, https://vimeo.com/142378753 (Accessed September 2019).

[34] Casey, B. J., Tottenham, N., Liston, C. and Durston, S. (2005). Imaging the developing brain: What have we learned about cognitive development? *Trend. Cognitive Sci.*, 9, No. 3, 104–110.

[35] Schwartz, B. and Sharpe, K. (2010). *Practical Wisdom.* Riverhead Books, New York.

[36] Blockley, D. I. (2010). The importance of being process. *Civ. Eng. Environ. Syst.*, 27, No. 3, 189–199.

[37] Blockley, D. I. and Godfrey, P. S. (2000). *Doing It Differently.* Thomas Telford, London.

[38] Hansford, M. (2014). Daring to be different: Sir John Armitt. *NCE* 18th February 2014, http://www.nce.co.uk/features/daring-to-be-different-sir-john-armitt/8659105.article (Accessed on June 2019).

[39] Godfrey, P. S. (2014). Capturing complexity in engineering systems, Seminar, University of Bath, Jan 2008 (Accessed on June 2014).

[40] Le Masurier, J., Blockley, D. I. and Muir Wood, D. (2006). An observational model for managing risk. *Procs. Insts. Civ. Engs. — Civ. Eng.*, 159, 35–40.

[41] Nicholson, D. P., Tse, C. M. and Penny, C. (1999). *The Observational Method in Ground Engineering: Principles and Applications.* Construction Industry Research and Information Association, London.

[42] Peck, R. B. (1969). Advantages and limitations of the observational method in applied soil mechanics. *Géotechnique*, 17, No. 2, 171–187.

Chapter 13

Systems Thinking — The Wider Context*

Bridges Are Built by People for People

As a civil engineer bridges fascinate me and I like the way lots of people use a bridge metaphor to describe social relationships. So, in this chapter I want to start with bridges to explore the link between hard physical and soft social systems.

Physical bridges are built by people for people. Bridge builders must work with and through others — they have to work in teams. Teams work effectively when relationships are good. Relationships connect people — as do bridges — so I will call them people bridges.

As individuals we build a people bridge whenever we make a new friend, or work with a new colleague. We maintain people bridges through our social lives. At home, at work or in a relationship some links may be relatively straightforward — making an acquaintance at a party. Others may be part of a complex situation — like deciding to change a job or whether to get married. Of course, the most difficult are the relationships between companies, nations, racial and religious communities. When people bridges are neglected relationships deteriorate.

---

*This chapter is an adapted and extended version of Chapter 7, "Bridges built by people for people: Processes for joined-up thinking", in *Bridges: The Science and Art of the World's Most Inspiring Structures*. Oxford University Press, Oxford.

The story of physical bridge building has many interwoven strands of artistic, technical, scientific and cultural development [1]. We can learn to read a physical bridge like a book. The letters of the book of a bridge are the constituents of materials such as the sand, cement and aggregate that makes concrete or the iron, carbon and other elements that make steel. Steel is a word as is concrete, timber and other materials such as the ground. The sentences are components such as steel plates and beams, the paragraphs are assemblies such as the towers or pylons of a suspension bridge. The chapters are entire sub-assemblies such as the foundations or a cable suspension system. The grammar of bridge structures is the way the components relate to each other and work together to form a successful whole. As we sift out the letters, words, sentences, paragraphs and chapters of the book of a bridge and delve into the grammar we can begin to appreciate their aesthetic, historic, social and engineering value.

Just as the book of physical bridges has many components so does the book of people bridges — but the book is very much more complex. Just a few examples will show how deeply embedded the bridge metaphor is in our thinking. Perhaps the most profound is as a link to the next life. Less seriously we refer to past experience as 'water under the bridge'. You may delay by saying 'I will cross that bridge when I get to it'. Feeling a bit philosophical — then life consists of 'many bridges to cross'. Options reduced — then you have 'burnt your bridges'. Bridges are embedded in popular music. Simon and Garfunkel sang 'Like a bridge over troubled waters'. Prime Minister of NSW, Australia, Jack Lang when opening the Sydney Harbour Bridge said, '... the bridge of understanding among the Australian people will yet be built'. We speak of building bridges of understanding between different nations, cultures and religious faiths. In short, the bridge as a link is almost as basic to the way we think as the ancient elements of earth, water, air and fire.

People bridges depend on how we are as individuals and how we express ourselves. We do that through the actions of our 'body language' and the words of our prose, poetry, drama, art and music. The 'chemistry' or grammar of the interactions between people *emerges* from those expressions and we attempt to capture it through prose, poetry and drama that is built on letters, words and sentences of natural language. Physical bridges can inspire — even be hailed as a form of public art. Art sustains people bridges. Music has letters, words, sentences, paragraphs, etc. as notes, chords, bars, phrases, periods, sections (a musical idea — verse, chorus, refrain) and movements that interact and emerge as the beauty of melody and rhythm of a composition.

Each of these kinds of expression use a hierarchy of symbols, materials and sounds and each have an impact. The effect is not a direct property of the individual letters, words, notes, chords or daubs of paint but *emerges* from the interactions and relationships between them. The interactions create the quality of content and performance of a people bridge.

Denis Noble in his book *The Music of Life* [2] uses music as a metaphor for the complexities of interactions at the various levels of an organism from genes to proteins, through cells, tissue, organs to the whole. He describes listening to a CD recording of his all-time favourite music, a Schubert piano trio. The digital code on the disc is transferred by the laser reader and converted into a signal which passes through the amplifiers and the loudspeakers and turned into sound waves that travel across the room to his ears. He describes how he is so caught by the emotion of the music of the moment that he cries.

So, **Learning Point No. 20** is that 'the book of a bridge is analogous to the book of a piece of art or an organism'. In an organism the letters are the complex molecules of DNA in genes, RNA and many others such as water and lipids which combine to make the 200 or so types of cells — the letters of the book of the human body [1]. The words of an organism are tissue such as flesh, muscle, bone and blood. The sentences are organs like heart, lungs, liver and brain. The paragraphs, sections and chapters of an organism are the complex interconnected combinations of organs that make up the musculoskeletal system, the circulatory system, the respiratory system, the endocrine system, the immune system and the reproductive system.

The processes in the various levels interact in complex ways to create emergent characteristics. For example, Noble describes his research into the mechanisms that cause the heart to beat. In one experiment the system is a muscle pacemaker cell with components which are protein molecules that channel the electrically charged ions to create the rhythmic activity. He describes how the components, the proteins, contribute to the behaviour of the system, the cell, but then the system feedback alters the behaviour of the components in a set of loopy processes.

Now you could be forgiven for thinking that physical bridges made of masonry, timber, steel and concrete, are much simpler than biological organisms and you would be right. Nevertheless, the same principles apply. The reason we think bridges are simpler is because scientific reductionism has worked very well for them. Indeed it has worked so well that we know that the science is true. But do we?

There is a question and it is a very big one. Why does science seem to work sometimes and not others? One obvious example is the science of food and its effect on our health. Scientific advice on what is good for the various aspects of our health is complex and changing, so many people are confused with the consequence that some just ignore it. Acid rain is another. Acid rain corrodes bridges although the damage to trees and other plants is more serious. Rain is naturally slightly acidic because carbon dioxide dissolves in rain drops to give carbonic acid. If there are strong emissions of sulphur dioxide and nitrogen oxide from industry then the rain can become even more acidic.

When there is a gap between what we know (the chemical reaction) and what we do (pollute) then unintended consequences can occur (trees killed). If we understand the problem properly and we have the collective will to communicate between different professional disciplines, we can act and build bridges over the gaps and deal with the problem. Once we recognize there is an issue we can take steps to deal with it — as long as we have time and the situation is indeed reversible. The gap between what we know and what we do was and is being bridged for the effects of acid rain — but not before some embarrassing and troublesome unintended consequences.

So, one reason why sometimes science works and sometimes it does not is that our understanding is always incomplete, messages change and sometimes conflict and we get unintended consequences from our decisions.

It is undeniable that reductionism is not only powerful and important but it is also irrefutable that it is not sufficient for dealing with really complex systems. The many successes in physical bridge building are impressive but we actually understand less than we sometimes think we do.

The scientific models are not absolutely true in all contexts rather they are contingently true in specific contexts. All physical man-made systems are built to work in the human context that gives them meaning — but that context is complex and embedded in the physical and natural world that we call the environment.

In earlier chapters (e.g. Chapter 8), I have used the word 'hard' to mean something that is definable and measurable and the word 'soft' to define something that is difficult to define. Traditional science is the science of hard systems based on reductionism. As Denis Noble says many of the latest developments in the science of the human body have been gained through the reductionist 'hard' science of biology and biochemistry. He argues that we now need to recognize that organisms are actually so complex that reductionism has to be set in a wider systems perspective.

As Noble says, we must not abandon reductionism. Rather we must see the emergent whole as well as the reductionist parts. Reductionism on its own has been enormously successful for hard systems but it reached its limit for some of our most difficult issues. It is unlikely ever to succeed for soft social systems.

Millions of people believe in perfection — mainly through their religion. But most realize it is impossible to achieve in practice. It is something to aspire to but there is always a margin between what we do and absolute perfection. We always fall short.

So what is realistic? Bridge builders have to be pragmatic — they have to be practical and build bridges that will work. Pragmatic systems thinking is natural for bridge engineers but this is not the same as the American philosophy of pragmatism which stresses that practical consequences determine meaning, truth and value [3]. Here we talking about practical problem solving (Chapter 12).

Bridge engineers have to be pragmatic about people bridges too as they form and reform teams for various projects. The hard practical fact is that no one can guarantee that disaster would not ever strike again. But whilst we may not be able to achieve perfection, we can strive to make future risks acceptably low.

But the situation is perverse. When disasters are rare then decision makers tend to push the issues to the bottom of their priority list. Then there is a danger of neglect. That is just what has happened to the maintenance of physical bridges. Almost all countries in the world have a decaying stock. The American Society of Civil Engineers said in 2003 that 27% of the 590,750 bridges in the USA are structurally deficient or functionally obsolete. Similar figures for 2015 show that 9.6% of the USA bridges are structurally deficient.

One of the UK Government's more successful initiatives has been that of 'Rethinking Construction' [4] — now rebranded as 'Constructing Excellence' [5]. Whilst at its best the UK construction industry is excellent and matches any in the world, too often it has underachieved with low profitability, too little investment in capital, research and development and in training. The industry is highly fragmented, trust is hard to build, and too many clients have been dissatisfied with the performance. The then Deputy Prime Minister invited a group of clients chaired by Sir John Egan to suggest improvements. They reported in 1998. Since then a number of changes have been made and achievements monitored. The five key drivers for improvement were committed leadership, focus on the customer, integrated processes and teams, a quality driven agenda and commitment to people: in other words, joined-up construction. Patrick Godfrey and I have promoted a systems thinking approach to these issues [6] but success is still very patchy.

Building bridges (hard or soft) is practical problem solving (Chapter 1). By problem solving I do not mean puzzle solving like a crossword or Sudoku but the creative challenge of identifying issues and meeting them to deliver a significant human need. At the same time we should recognize that the flip side of a problem is an opportunity — so bridge building is also about finding and taking opportunities.

One way of beginning to understand differently is through a seemingly trivial problem (first suggested by Peter Senge in his wonderful book on systems thinking [7]) — 'What do you actually do when you turn on a tap to fill a glass of water?'

Before reading on please pause a moment. Think about all of the important steps in the process of turning on the tap and filling the glass. Now, if you can, write them down and do this before reading on.

Perhaps you wrote a list or drew a diagram something like Figure 13.1(a). If so, you are not unusual because most of us are trained to think in straight lines.

But if you reflect for a moment turning on a tap is not a linear process, it is a loopy one. As you turn the tap you are constantly watching the water level and adjusting the water flow until the level gets to what you want. So I have redrawn the stages as processes (see Figure 13.1(b)) where the arrows represent the way one process of change influences another. The result is Figure 13.1(c).

We start with a need — the desired water level. We perceive a gap between the water level and the one we want so we open the tap by adjusting its position. As a consequence water flows and the water level changes. We see a new gap and adjust the tap position again. We go round and round this loop until we get the level we want and then we turn the tap off. Simple! Yes, but more complex than perhaps it first appears — as anyone who has tried to design a robot to do this kind of task will tell you.

In general, solving a problem is loopy and the stages are sub-processes as I have drawn in Figure 13.1(d). I have used the present participle, the 'ing' form, to capture the processes as active 'doing'. Each of the five stages needs to be managed to success if we are to manage the original

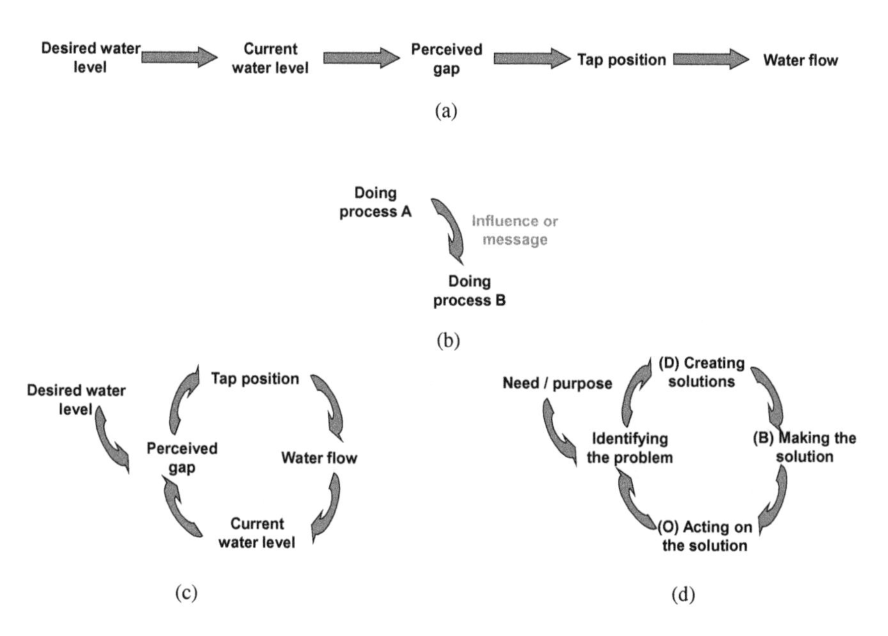

Figure 13.1. Processes for drinking water.

problem to success. The sub-processes are: understanding the need, identifying the problem, creating solutions, making the solution and acting on the solution. As we go around this loop time moves on and the situation changes. The result is that when we reach the end of the loop we now have a new state of affairs and new problems and altered (may be only slightly) needs. So we have to start going around the loop yet again. The loop is a continuous spiral through time.

But there is more. Each sub-process will have sub-sub-processes — indeed they are *all* parts and wholes — holons. So there are process holons at every level of definition all going through spirals. From the top process holon right down to every detailed process holon there are continuous interdependent problem-solving spirals.

Let us look at the five stages of Figures 13.1(c) and 13.1(d) in a bit more detail. The first is the driver of the spiral cycle — need or purpose. One way of finding this is to simply ask why? Why are we doing this? Why do we pour water into our glass? — We need a drink. Why are we building a bridge? — We need to cross the river. Need and purpose are the potential or voltage that drives the current of change around the loop. It is the gravity force that makes the glass you are holding fall to the floor if you let go.

The second stage follows directly on — identifying the nature of your problem. You can think of it as part of the first stage of asking question why. What level of water do you need in the glass? Perhaps a ferry is all that is needed to cross a river? Do you need a footbridge or a highway bridge? Just what are you trying to achieve in solving this problem? What are your aims and objectives?

The third stage is deciding what to do about your problem i.e. how far should you turn the water tap. More generally this stage is about creating possible solutions to meet our needs or solve our problems and deciding on the criteria for choosing one of them. We will call this D for *designing.* For a physical bridge we must decide on the form, materials and detailed structure. Designing is an 'opening out' creative process where ideas are suggested and developed. Note that designing is more than simply creating something that looks good — aesthetic of form. It is also about creating something that works. But more than that it is about creating something that works elegantly — an aesthetic of function — and also about choosing the criteria of what is good — deciding why one solution is better than another. Those criteria have to reflect a whole range of needs from function and form to cost and sustainability. So designing is an evaluative process of deciding.

The fourth stage is that of doing something or of something happening — like the flow of water through the tap. The 'state of affairs' changes — usually irretrievably — there is no going back. We will call this (B) for *building* — making the solution we have designed happen. We actually change our reality — we implement our solution — we build the bridge.

The fifth stage is acting on the solution — we will call this (O) for *operating* or *using*. We work in the changed world with the changes we have made and we operate our solution. We monitor the changing water level. We use the bridge we have built. Now we have a whole new set of problems. For example, we have continually to maintain the bridge and, at the end of its life, we have to demolish it.

How do we use our understanding of these five problem-solving stages to create a successful bridge? I maintain that we have three interdependent strategic needs or requirements to satisfy. If we can meet these needs successfully we will create a successful bridge. We need *firm foundations* on which we build a *strong structure* that will *work effectively*. We need to find ways to design, build, operate and integrate them. They are our 'high-level' processes. In other words, success in the three process holons to meet these needs are together necessary and sufficient for the success of the whole bridge. But because the three processes are interdependent they need a single integrating process. So, our top-level need and purpose is to make sure that they are integrated — it is a kind of meta-purpose, i.e. one which surrounds the more obvious need and purpose of building a bridge to bridge a gap. So, at this top level the need/purpose and the process of identifying the problem in Figure 13.1(d) have become the single integrating process. I have put this at the heart of the three interacting problem-solving loops in Figure 13.2.

Let us examine each of the three interdependent needs more closely. First are the foundations. Foundations hold the rest of a physical bridge in place. But how important are the foundations for a good people bridge? One word captures the idea of a good foundation — purpose — the underpinning purpose of the bridge itself. For people bridges a good foundation is a firm foundation for life. It holds the rest of our lives in place — it

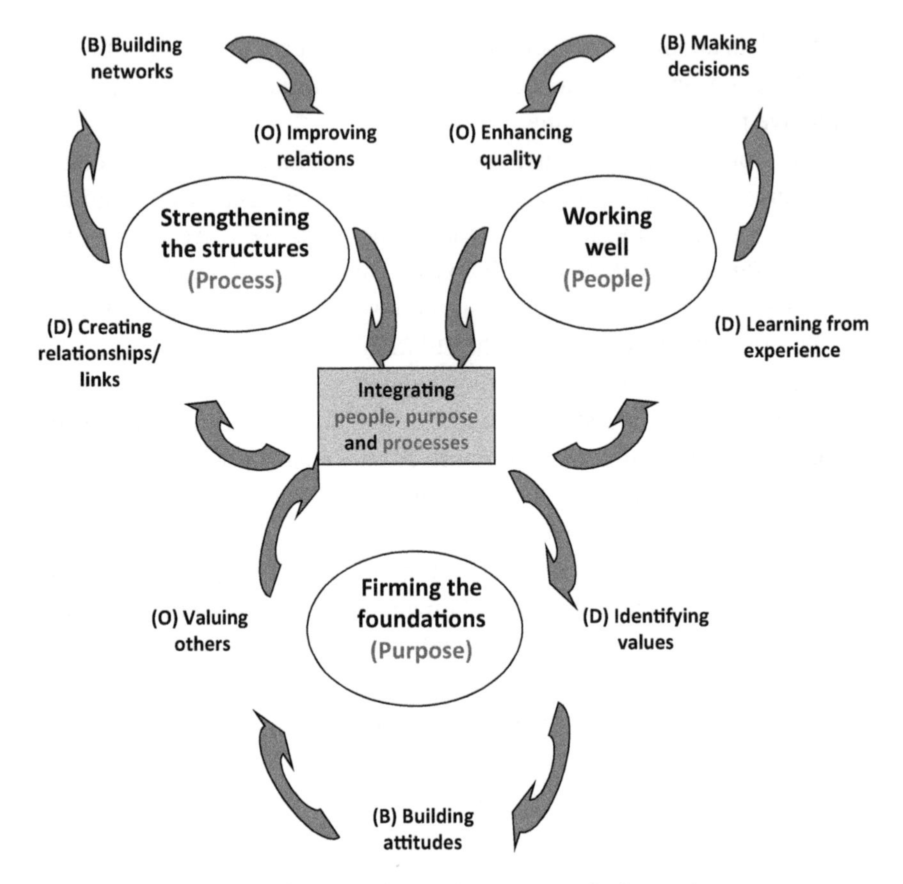

Figure 13.2. Three interacting processes for integration.

underpins all that we do — it gives us a sense of place and purpose. It is in very large part provided by our education and upbringing. We *design* people bridges through our common values. From our values we *build* attitudes and through them we learn to live (*operate*) our lives through valuing others as we grow. Firm foundations help us cope. They help with life's challenges and enable us to be robust. With weak foundations we succumb and we lose our way. In short, firming the foundations is about laying down the basis of the way we live with others to provide a quality of life. This hopefully includes, for most of us, religious and racial tolerance and the many other things that lead, ultimately, to our own self-fulfilment and happiness. Groups, clubs, companies, charities, governments and large multinationals all need solid foundations too — common purpose developed through good leadership. It is no coincidence that many organizations are called foundations when their whole basis is based on a legacy or donation. Firm foundations provide organizations with a sense of purpose. Organizations of all kinds flourish if the basics are right with visionary clarity of purpose.

The second requirement for a good bridge is a strong structure. Bridges are links — they join things together to form a higher-level structure — for example, as part of the infrastructure of a region. Just as we can span a physical divide, such as a river, with a bridge of steel, so we can span deep intellectual, emotional, religious and cultural divides, such as self-interest, fear and intolerance, with bridges of understanding. One word captures this idea — process. In a physical bridge, the process is the flow of internal forces through the physical set of interconnected pieces of material which is the structure. So, the structure is more than a set of 'things', it is the framework for a set of interacting processes.

In a people bridge, the structure is a set of people and community relationships which serve to communicate messages of information and meaning. In other words, a structure is the form of a process. Without structure nothing can work — everything must have structure — the only question is whether it is a good structure, i.e. fit for purpose. That is not to suggest that bridges of understanding are easy to create — sadly and all too frequently the opposite is true. Some gaps are so wide and the issues so tough that they can never be solved — only resolved. Violent conflict, in Syria for example, has cut so deeply into communities that it will take many years, even decades, to find peace and reconciliation. Bridge building in these situations requires long-term 'joined-up' thinking to create links at every level from individuals through groups to governments. Just as the elements of a physical bridge must relate well to each other so must people and communities. Tragic human stories, including child murders such as that of Victoria Climbié [8], often show a remarkable lack of joining-up.

The third requirement is that bridges must work well. One word captures this idea and that is people. This loop represents the performance of the physical bridge (which, although it is a physical object, is perceived, understood and responded to by people) and the performance of people and communities through living their daily lives. Ultimately all bridges are there to work for people.

What is the nature of the process needed to integrate or join-up these three processes of firming the foundations, strengthening the structure and working well?

I define an integrated fully joined-up system as one where we get the right *information* to the right *people* at the right *time* for the right *purpose*, in the right *form* and in the right *way*. We can unpick this rather long definition using *what, who, when, why, where* and *how*.

Think of the information as answers to questions starting with the word *what* — simply what is the information we need and may want to transmit? Think of the people as the answers to the questions *who* — simply who is involved? Think of the time as answers to the questions *when* — simply when is it needed. Think of the purpose as answers to the questions *why* — simply why is this needed? Think of form as answers to the questions *where* — simply what form should the information be expressed and what assumptions should be made about its context. Finally, think of the way as answers to the questions *how* — simply how should the information be transformed. So, the definition just boils down to asking six questions *what, who, when, why, where* and *how*.

But this is all very easy to say — but not so easy to do especially across all processes in the entire system. If there is a single deficiency in any of these requirements then there will be a lack of joining-up. All it takes is a message, or piece of information, that does not get sent or received, or is poorly formulated, incomplete, misleading, or is without adequate justification.

Joining-up is not just about the flow of information between people it is also about the flow of a physical material. Think of the flow of water around your house in network of pipes or from the waterworks along a network of pipes to your tap. Think of the electricity flowing around the ring main of your house or on the national grid. Think of the flow of forces through a bridge structure. Any lack of joining-up in these systems can mean no water at the tap, no electricity at the wall socket or a bridge falling down because the forces become unstable. So, we are able to create successful joined-up systems — but they are ones that have yielded to a scientific reductionist approach.

At first sight these two interpretations of joining up are very different — one about people and one about physical things. But they are both about potential driving change or *vice versa* — so we can join them up conceptually. That way we can see better how people are connected to their environment. We need to do it to minimize the unintended consequences that are so often at the root of why things go wrong.

Change is flow — continuous movement. When we think about change, no matter the timescale, we look for reasons or causes and effects. The idea of cause and effect is ingrained in our practical everyday thinking but has been a concept that has troubled philosophers ever since Aristotle. Common sense tells us that if a wine glass falls to the floor (effect) it is because someone or something dropped or pushed it (cause). But the effect may then become a cause of a different effect (wine staining the carpet). So causes and effects are not simply distinguishable. This kind of common sense does not help when we postulate the fundamentals of our existence — we humans exist (effect) because God made us (cause). Such statements are a matter of faith.

In Figure 13.3, I have attempted to use this way of thinking to capture the story of the wobbly Millennium Bridge over the Thames in London [1].

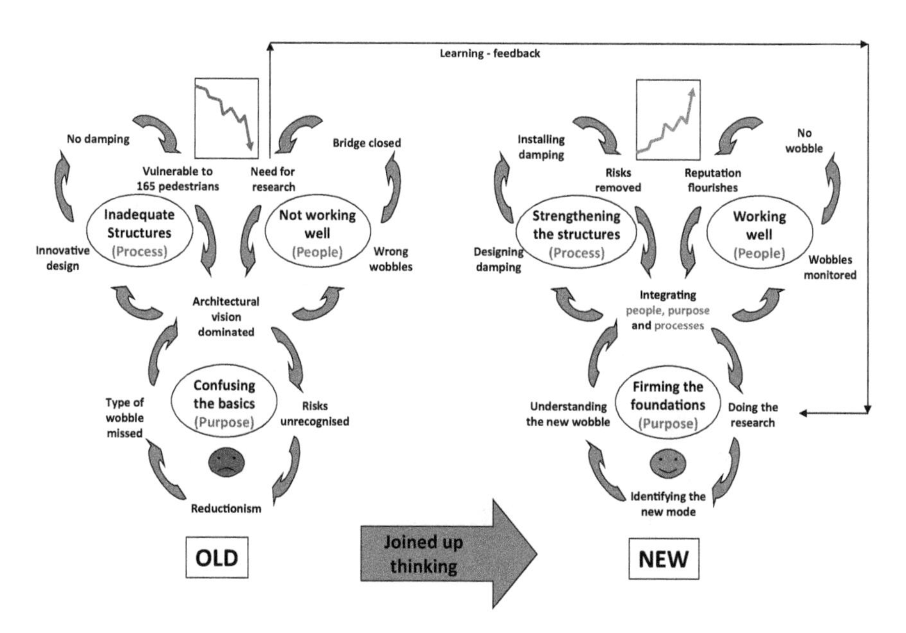

Figure 13.3. Learning from the London Millennium Bridge.

The left-hand loops represent the original story. The arrow link between the 'Need for research' and 'Doing the research' is the learning — feedback. It is a step change in the relationships between the processes (from 'Old' on the left to 'New' on the right of Figure 13.3) forced by the closure of the bridge. The right-hand loops are the new situation after the research in synchronous lateral excitation.

Figure 13.4 refers to the abuse of an innocent little girl. When I proposed to include this diagram in my book on Bridges [1] my publishers said that readers would not be able to relate the impersonal analysis of a river bridge with the story of an innocent girl so it was edited out. Now I maintain that my analysis shows patterns of a lack of joined-up thinking that merits attention — even if the subject matter is so different. Both are actually human stories but of course the Victoria Climbie case pulls at the heart strings. It would be good to report that lessons had been learned but unfortunately similar cases keep occurring such as Baby P [9] and Daniel Pelka [10] where the authorities had many chances to uncover the abuse.

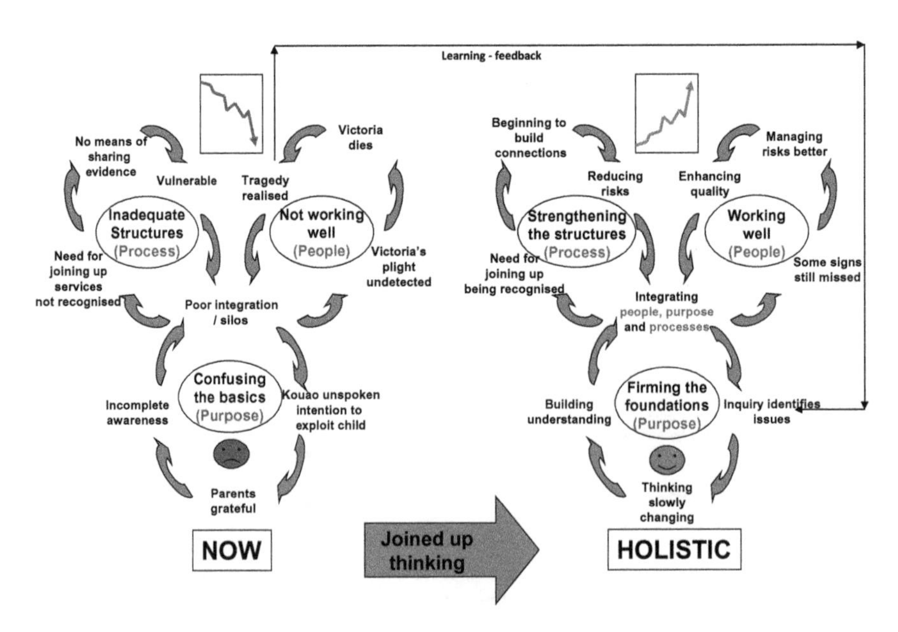

Figure 13.4. Learning from the tragic case of Victoria Climbié [8].

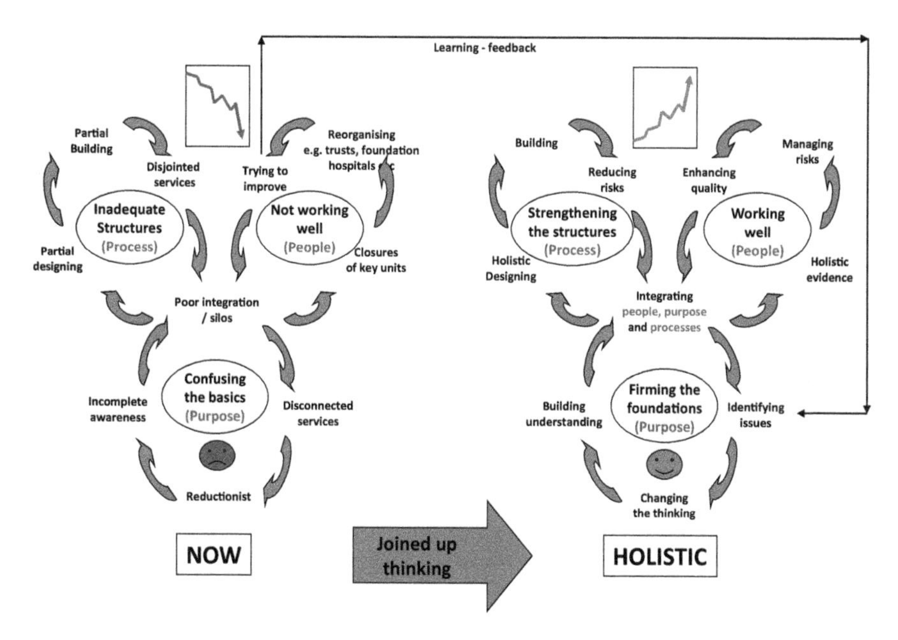

Figure 13.5. Joined-up thinking for the NHS.

Figure 13.5 refers to the UK National Health Service. This is such a massively complex organization but one which enjoys the support of the vast majority of the UK population. In my direct experience it can work well though one can always see the potential for improvements. The national problem is that some underperforming hospitals receive sensationalist publicity when something serous goes wrong. The NHS is also a 'political-football' and so is constantly being asked to change in quite radical ways before the effects of the previous changes have been evaluated. Figure 13.5 is obviously simplified to give an overview of how systems thinking can help. Again the central point is joined-up thinking by integrating people, purpose and processes.

My final example concerns our limited ability to predict the future (Figure 13.6). That underpins why there has been so much controversy surrounding the reports of the Intergovernmental Panel on Climate Change [11] and what needs to be done.

One consequence is that many people shrug their shoulders and say that is up to others to do what is necessary. It is easy to think that your own contribution is so small as to be insignificant in the totality of this enormous issue.

Clearly, on some issues, it is the specialists who have to act. For example, it is prudent to assume that physical bridges, like the rest of the built environment, will have to cope with more extreme weather events. Bridge builders have a responsibility to cope by developing even better methods of risk management.

Likewise, our political and business leaders would be prudent to assume more turbulent times ahead. Again this calls for better long-term risk management.

If climate change is as urgent as the specialists tell us then we do not have much time. A hard systems analogy may help. In an electrical circuit resistance is the dissipation of energy. War and internal conflict are all dissipating resources that ought to be invested in tacking climate change. So war and conflict are resistance in the process of tackling climate change. Conflict reduces our capacity to act together.

It is clear that if we are to act collectively on the timescale needed, we will need new cooperation through new people bridges. But at all levels,

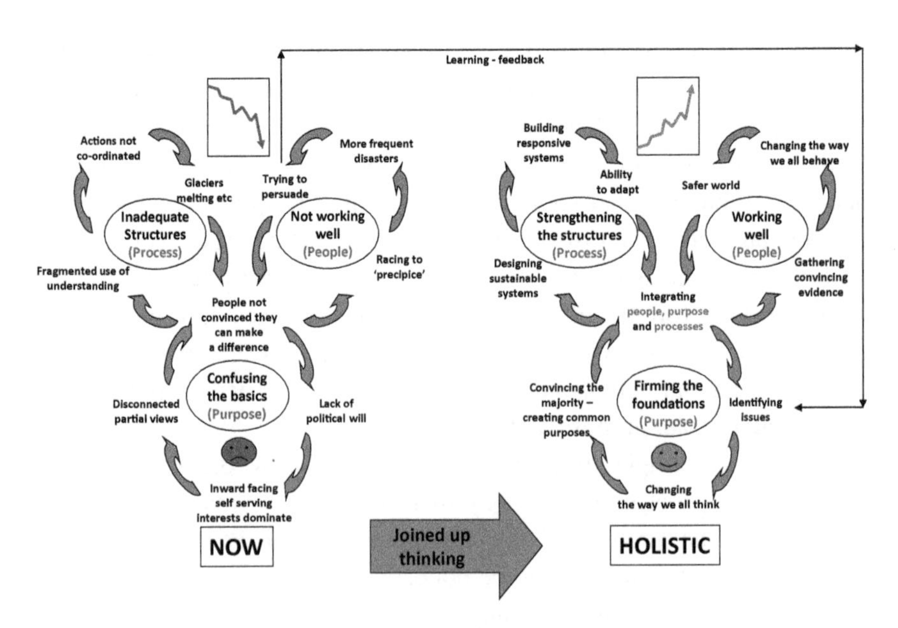

Figure 13.6. Joined-up thinking for climate change.

from the letters to the book of the whole world, it is the people bridges that are so crucial and the hardest to build. Whatever we do, at whatever level we operate, unintended consequences will occur. Our best strategy is to absorb current reductionist thinking into a system thinking evolutionary observational approach. As we work at each level at all of the issues we face we must be mindful of our effects at other levels. What each of us does might seem small in comparison to the whole but the emergent properties of what we all do together are what we should be focusing on.

Dealing with climate change urgently requires a shared clarity of vision and purpose at all levels with a sense of total integration. It requires us to expect the unexpected and therefore to be prepared to change and adapt. Then we have to integrate and interpret what is happening. We have to get better at valuing real evidence. Most of us need help to sort out the statistical 'wheat from the chaff'. The media have an important ethical role here within their commercial purpose. They must help people to understand statistical evidence rather than, as is all too often the case, to mislead with implications drawn from anecdotal stories. All of us have then to plan what to do at our many and various levels of being and we need to do it for all foreseeable outcomes. But we have to do that whilst expecting some complete surprises. As we do in everyday life we deal with surprises as best we can.

The slogan 'Save the Planet' is the most misleading ever. The planet will survive — the question is whether the human race will. Bridge building has much to offer.

References

[1] Blockley, D. I. (2010). *Bridges*. Oxford University Press, Oxford.
[2] Noble, D. (2008). *The Music of Life*. Oxford University Press, Oxford.
[3] Dewey, J. (1930). *The Quest for Certainty*. George Allen & Unwin, London.
[4] Department of the Environment Transport and the Regions (1998). *Rethinking Construction*. London.
[5] Constructing Excellence (2019). http://www.constructingexcellence.org.uk/ (Accessed March 2014).
[6] Blockley, D. I. and Godfrey, P. S. (2000). *Doing It Differently*. Thomas Telford, London.

[7] Senge, P. M. (1990). *The Fifth Discipline — The Art and Practice of the Learning Organisation*. Century Business, London.

[8] Department of Health and The Home Office (2003). *The Victoria Climbié Inquiry*. Report of an Inquiry by Lord Laming, Cm 5730, January.

[9] Haringey Local Safeguarding Board. *Serious Case Review: Baby Peter*, https://www.haringeylscb.org/sites/haringeylscb/files/executive_summary_peter_final.pdf (Accessed September 2019).

[10] Coventry Safeguarding Children Board. *Serious Case Review: Daniel Pelka*, https://www.lgiu.org.uk/wp-content/uploads/2013/10/Daniel-Pelka-Serious-Case-Review-Coventry-LSCB.pdf (Accessed September 2019).

[11] Intergovernmental Panel on Climate Change, http://www.ipcc.ch/ (Accessed March 2014).

Glossary

Word	Description
Capacitance	Holding electrical *charge* or more generally storing flow to create potential.
Charge	A basic property of matter that creates electric and magnetic forces and is positive or negative.
Complex system	Complex *systems* are difficult to describe and predict because they have many interconnected parts with *emergent* properties. They may, under certain conditions, behave in a chaotic way. A complex *system* is not just *complicated* but may be incomplete with *emergent* properties from *interdependencies* that are unknown and unforeseen.
Complicated system	A complicated *system* may contain lots of elaborately interconnected sub-*systems* or *holons*. To the non-expert complicated *systems* may seem complex but to an expert each of the sub*systems* may be *tame*, i.e. understandable and predictable.

Word	Description
Dependability	Being dependable, trustworthy and reliable. Practitioners require information that is *sufficiently* dependable for the decisions they make. Truth is *sufficient* but not *necessary* for dependability, e.g. Newton's laws are not strictly true in that propositions can be deduced that do not correspond with the facts for bodies travelling at velocities near the speed of light.
Deterministic	The philosophy that all events are totally determined by previous events with no *uncertainty.*
Duty of care	A duty of care is an obligation to not act negligently under the Law of Tort.
Emergence	A property of a whole that emerges from the interactions of the parts. It is the reason why a whole is more than the sum of its parts.
Energy — potential, kinetic, strain	A capacity for work — potential is due to position, kinetic is due to movement and strain is potential in a deformed material.
Entropy	The loss of available energy in a heat engine or loss of information in a signal — an increase in disorder.
Fuzziness	One of the chief structural elements of *uncertainty* which is vagueness and imprecision.
Hard System	Any physical system such as a building structure or railway line which involves action and reaction, c.f. *soft systems* that involve people. Hard systems do not depend on who is interacting with the system, they are assumed to be independent of the observer and hence are the same for all of us.
Holistic	The idea that the whole is more than the sum of its parts.
Holon	Something which is both a part and a whole at the same time.
Incompleteness	That which we do not know.

Word	Description
Induction	A process in which electricity or magnetism produces electricity or magnetism in another body without any physical contact. More generally storing potential to create flow.
Interdependence	Where two *processes* are mutually dependent.
Italian flag	Used as a colourful indicator of an interval probability. For example, in the statement 'the probability of A is $p(A) = [0.4, 0.8]$ then the interval 0–0.4 is coloured green (representing the evidence for success), 0.4–0.8 is white (representing what we do not know and 0.8–1 is coloured red (representing evidence for failure). Green, white and red are the colours of the Italian flag.
Logos	Rational and pragmatic reasoning about facts and external realities — the kind of reasoning we use to get something done but says little about religion, emotions and the meaning and purpose of life.
Model	A representation of something.
Mythos	Understanding derived from storytelling, often mystical, religious, emotional, and rooted in the subconscious mind.
Necessity	The state of being required to be done or achieved, imperative or indispensable.
Newtonian physics	Alternatively known as classical mechanics and based on the laws of motion formulated by Sir Isaac Newton (1642–1726).
Object	An object is anything that, by being to some degree stable or coherent, may be apprehended, perceived, understood or conceived through thought or action.
Objective	1. A precise statement of an intended outcome — it is a target for success in a *process*. 2. Knowledge and information that is not *subjective*, i.e. exists outside the mind of anyone individual and is available to everyone.

Word	Description
Phronesis	Aristotle's notion of *practical wisdom* and prudence.
Potential	Capable of becoming, voltage in electricity and velocity in mechanics.
Practical rigour	Practical rigour is different from theoretical rigour. Practice requires a use of rational judgement that transcends strict logical rules to deal with *uncertainty*. Engineers use *appropriate models*, find *dependable* evidence and with a *vision* of what is needed create solutions that work. Practical rigour requires practical foresight and analysis of all hindsight. The possible unintended consequences of human action are legion and the rigour of practice is about anticipating and managing them.
Practical wisdom	A *quality* of discerning and judging — a way of looking at things with an ability to see the world in a coherent picture. The way a person constructs the world in which they operate, which in engineering is to do with having *appropriate models* to fit the situation. Practical wisdom implies practical rigour which in turn implies practical intelligence which implies practical experience.
Process	That which is done to change a state of the world. A basic tool for describing a *system*. It is a way of getting from where you are now to where you want to be.
Process model	An arrangement of processes such that the lower level processes work together to achieve the *purpose* of the higher process.
Purpose	The reason for doing something. The *intention* or *objective*. The result or effect that is intended.
Quality	Fitness for *purpose* and degree of excellence at the same time.
Randomness	Lack of a specific pattern in some data.

Word	Description
Reading an object such as a bridge	Observing, understanding and interpreting an object to give it meaning.
Reductionism	The idea that a system can be completely understood by understanding its parts or components, i.e. the whole is merely the sum of its parts.
Reflective practice	Acting with *practical wisdom* and *rigour* — the way professionals do their work. They perceive the world, they reflect upon it and they act. They do it with rigour, wisdom and foresight.
Resilience	The ability of a *system* to withstand or recover quickly from challenging conditions; to respond by detecting, preventing and, if *necessary*, handling disruptive challenges.
Resonance	Where the frequency of a stimulus is close to the natural vibration frequency causing very large vibrations.
Robustness	Sturdy, *resilient*, durable and hard wearing. A *system* that is robust is not vulnerable.
Soft System	Human and social *systems* involving people. Our ability to *model* them and hence predict them dependably is very poor. Soft systems comprise action, reaction and *intention*, c.f. *hard systems.*
STEM	Science, Technology, Engineering and Mathematics.
Sufficiency	The state of being enough, being adequate for what is required to be done or achieved. A sufficient condition is one that, if fulfilled, is adequate.
Sustainability	Keeping something going over time — keeping from failing or enduring without giving way.
Synergy	Synergy occurs when parts combine to produce a total effect that is greater than the sum of the individual parts.

Word	Description
System	An overused word with many meanings and uses. Used loosely it just refers to any group of connected *objects*. More precisely as used in systems theory it is *a complex* whole with a set of interacting parts as a connected network or mechanism which has *emergent* properties that explain why the whole can be more than the sum of its parts.
Systems thinking	A way or philosophy of approach to problem solving that values both parts and wholes, i.e. one that combines reductionism with holism.
Uncertainty	Absence of precise and complete knowledge.
Value	1. The regard, merit, importance or worth given to something. 2. Material or monetary worth. 3. The magnitude of a mathematical variable, e.g. the value of x is 2 or $x = 2$.
Vulnerability	A property of a *system* where small damage can cause disproportionate consequences.
Wicked problem	Problems and situations that do not seem to yield to easy solutions — contrasted with *tame*.
Work	A transfer of energy — exertion or effort — as a force moves a distance. It is measured as horsepower or joules as force times distance so one Newton of force moved one metre is one Joule.
Worldview	The way we look at the world, our point of view. We attribute meaning to things by interpreting it in the light of our education and experience.

Index

S

Lightning Source UK Ltd.
Milton Keynes UK
UKHW020647170120
357139UK00001B/32

9 781786 347626